Drowning in the Lake
while Embracing the Reflection of the Moon

Drowning in the Lake
while Embracing the
Reflection of the Moon

Seeking the Source of Underlying Reality
while Clinging to Delusional Convictions

ROBERT GRAY HOWARD

Drowning in the Lake while Embracing the Reflection of the Moon
Seeking the Source of Underlying Reality while Clinging to Delusional Convictions

iUniverse books may be ordered through booksellers or by contacting:

iUniverse
1663 Liberty Drive
Bloomington, IN 47403
www.iuniverse.com
1-800-Authors (1-800-288-4677)

ISBN: 978-1-4917-6223-3 (sc)
ISBN: 978-1-4917-6224-0 (e)

Library of Congress Control Number: 2015903416

Print information available on the last page.

iUniverse rev. date: 03/25/2015

Professional Researcher Preface

The professional researcher, with scientific, religious or other world view, must achieve results which are usually presented to others in writing or in speech. This book aids the research and the presentation as follows.

- ✓ Provides more than sixty assumptions and hypotheses from which you can choose those to verify or falsify. Then you can publish your conclusions.
- ✓ Introduces all the recommendations you will need to design an experiment or a research project which will increase your prestige.
- ✓ Raises the ancient questions about God, Tao, and the underlying source of reality to aid your planning, your researching, and your publication program. This stimulates your curiosity.
- ✓ Presents the detailed reality of how three time dimensions emerge from the mind and consciousness.
- ✓ Provides exact scientific answers to the problem of identifying the information within the communications from the source of reality. This alone is easily worth four hours of reading.
- ✓ Suggests research guidance for discovering the communication channels from the underlying reality to kick off research for your numerous insightful publications.
- ✓ This book is priced low so the cost/benefit will pay off in ideas to get massive grants to fund years of your research.
- ✓ The best news is the large collection of new and exact descriptions of many realities.
- ✓ Secret revealed: how to remove noise from communications

Preface

The recipe for this book in given in the Foreword. Read the Foreword to understand the motivation and the goals of this book.

Savor the full flavor of this research into the source of underlying reality. Translate the title of this book into *How to Communicate with God,* which is the Christian-Jewish-Muslim world view.

To smell the full aroma of human delusional convictions, investigate the meaning of this ancient Chinese experiment with several levels of mind:

The all powerful and invincible Dragon
sails through the enchanting clouds
which obscure Heaven.
The arrogant, cunning and wise Tiger
sneaks through the mists
which hide Earth.
Heaven and Earth totally control Man
until the clouds and mists disappear.
Then a mosquito and a cockroach
are all that remain.

To test the enigmatic essence of the source of reality, read the end of the Diamond Sutra where a monk recalled the Buddha's description of the ineluctable essence of the source of reality.

Thus have I heard:

"This is the correct view when contemplating our conditioned existence in this fleeting world:

Life is like a fault of vision,
Like a tiny shining drop of morning dew, a bubble floating in a stream;
Like a fleeting Autumn cloud, a flash of summer lightening,
Like a shimmering star, a flickering lamp,
Like an illusion, a mock show, a phantom, or a dream.
So is all conditioned existence to be seen."

Table of Contents

Foreword

I have researched computer software; naval machinery, high energy physics, piping stress, solar energy, and marriage. As an engineer, I designed high rise renovations of heating, ventilating and air conditioning. I founded Sukhavati School in 1997. I was married and raised a daughter. I earned a BSE in Mechanical Engineering, a Master of Administrative Science, a Ph. D. in physics, three degrees in Japanese tea ceremony, and a state of mind in Zen Buddhist training. These disciplines focused on analyzing the causes and probabilities of complex environments.

Therefore, the I am entitled to raise the topics of

> ➢ identifying underlying reality
> ➢ human drive to power as the underlying reality of war and totalitarian government,
> ➢ the massive power of the human mind including extra sensory perception, and
> ➢ methods of communicating with the underlying reality.

Examining and questioning myriad processes led me to doubt the consensus reality. The earth exists on many levels of reality depending on how its elements are analyzed. The ultimate question is, "Is there a final underlying reality and how can I communicate with one or more underlying realities?"

The reality of civilized, modern large groups of humans

I am a serious student of the Way of life proposed by Buddha 2500 years ago. There is a short summary of some principles of the Way in *The Foundations of Mindfulness*. (Buddha, 500 BC) A part of the process of integrating the Way into my life has been purification. Purification consists of removing layers of elaborations that are considered civilized, modern life.

A few people notice that what is sensed is often found to be a deception. Some people conclude that ideas met in sources of information are often found to be false or purposefully presented lies. Careful scientific observations have uncovered underlying reasons for the behavior of living things. These observations began in prehistoric times. Focused research in science and religion have clarified these observations and answered problems related to cause and effect.

What would one find out if these researches were to extend to the final layer of underlying reality, the first source level of reality? This is the question this book seeks to answer. Many descriptions of reality are presented. Many methods of researching the answers are suggested.

Let us imagine how one can initiate a fruitful investigation into this question of reality. One begins by removing all distractions such as sense information, preoccupations with ownership, making one's self more important, and the drive for power. One begins by stopping the thinking about such distractions, preoccupations and so on. One ends by gaining knowledge of the ultimate reality. This knowledge is gained after distilling one's existence until all that remains is unlimited Being.

One can begin such research by reducing the total distractions and preoccupations into several goals and achieving one goal at a time. Finally, the question can be answered.

The life of an ideal Buddhist monk removes the distractions and preoccupations one at a time until there are no elaborations of sense data or thinking or knowing.

Consider removing the elaboration of tasting. The basic need is to eat nourishment with all the components necessary for health. The reality is about 30,000 kinds of food at the end of a complex supply chain extending thousands of miles in refrigerated airplanes, trucks, and warehouses. There are hundreds of different classes of restaurants. There are cooking schools. There are electric machines for cooking and preparing foods at home and in huge industrial enterprises. This is summarized by John Lanchester in "Shut Up and Eat." (Lanchester, 2014b)

There are mental self-deceptions such as with wines which sell for thousands of dollars for a bottle to experience its complex tastes. The deception arises in the denial that the basis of alcoholic drugs is to modify the feelings of the drinker; to get high, stupid or drunk.

Ponder removing the elaboration of sounds. There is a music industry with thousands of different kinds of music, hundreds of kinds of music making instruments including electronics. There are colleges where music is taught; making music with instruments and with electronics. There are electronic machines for recording and for playing back sounds. There is an industry that removes sound from buildings and from cars. Reality is distorted in the sounds of radios and the sound of language.

Investigate the elaboration of sights. There are movies, television, smart-phones, landscapes, and graphic symbols. There are electronics to produce deception in movies, smart-phones, and television. Movies, smart-phones, and television present a distorted reality that is often not based in the actual reality. Gadgets are the preferred means of entertainment, not a gathering of information necessary to survival. The trend is for sights to be those of an individual person not those of a community. This trend removes some of the elaborations at the cost of the treasure of community membership.

Clothing has changed from being a means of protection from the weather to being a visual communication. It has become a means of visual expression; communication through appearance. Clothing has a million expressions, brands, symbolism, international trade, and massive amounts of money transfer. Part of the sensation of clothing is touching, sense data. The touch of the clothing fabric, as it is worn, is another distraction. Reality is distorted so people are driven by obsession to wear certain clothes. There are infinite deceptions built into clothing. There is self-deception that one has changed one's personality due to wearing certain clothes; costumes. Other people, seeing the clothes also think one has changed personalities.

The deception of clothing is amplified in venues made to enhance one's emotions like a stage show or a huge dance floor. The combination of high energy sounds, bizarre costume sights, multiple aromas, dozens of drugs, and the pollution of the psychic field stimulate the senses beyond the capacity of the nervous system. This often leads to mental disorders: drunkenness, hallucinations, psychosis, and death.

Thinking is elaborated into a thousand fields of learning. It is stimulated by the senses which are distorted by elaborations.

Thousands of colleges and schools teach ways of thinking, journalism, public image, and stage acting. There is the mental disorder, accepted as normal, of thinking of the following examples as more real than sensed events: use of gadgets such as phones, television shows, smart-phone propaganda, and advertising,. The deception arises in preferring to talk to the smart-phone rather than to a person two feet away. The self–deception arises in gloating over knowing a scientific fact that no one else knows; of having a college degree no one else has, in believing that the advertisement is true when it promises you will be being happy after spending $1000 to go 1000 miles away and lying on the beach. There are many analyses of self-deception, even in books, because it is so common.

There is self deception of being in close communication with a multitude of people and external information by using phones. But the only reality is the communication with a gadget. The rare face to face communication is disappearing and is less articulate, lacking emotional content when it happens. The gadgets replace the reality of humans, trees, and emotional interactions. This is removing the elaboration of face to face communications by increasing the overall complexity of the technology underlying civilized modern groups.

Recall that feelings are a powerful part of thinking. There are about five thousand mood altering drugs. There are accepted dangerous drugs such as tobacco, alcohol, illegal drugs, and medical drugs. Thus, there is the deception of satisfaction after taking a drug. Feelings are elaborated into drug events that cause extreme mood changes.

Think of the elaboration of transportation which has supplanted walking and running. Each mode of transport has a huge industry which did not exist a hundred years ago. Now people run and walk as a separate exercise on a machine at a gym, not to arrive at a useful location.

Each of the senses has endless corporations producing deception of the senses. Each of the corporations and government regulation bodies has power mad people as chiefs who are mainly interested in each one's personal benefits to be gained from deceiving the senses of the customers. The drive for power is acted out behind a mask of providing a better life for the customers and other public relation propaganda.

There is the elaboration of the need for food, clothing and shelter into needing money income. This leads to the preoccupation of job training and professional training. This is supplied by massive worldwide school systems.

When one has money income, one elaborates the need for material things and experiences. This is composed of ownership, the drive for power to get things, to get experiences, and to feel pride [or perhaps jealousy] because of ownership.

There is an archetype in the human mind to crave gossip. This has been elaborated into the gossip called news. This propaganda stimulates common mental hindrances:

Doubting the truth and believing lies,
Hating and ill will,
Laying about in torpor and sloth,
Hurry, worry, restlessness, wasted busy-ness while going no where, and worry,
Craving and clinging to things, feelings, and experiences.

One begins the researches into the absolute final source level of reality by removing all distractions such as sense information, preoccupations and so on

There is a way to initiate a fruitful investigation into this question of reality. This book provides the bare guidance to the achievement of removing all the layers that hide the underlying reality. Many descriptions of reality are presented. Many methods of researching the answers are suggested. Persisting in this removal process will yield to you the view of the world without deception.

A monk can purify himself to perceive the world as it is without the deceptions and distractions

The monk has to purify out all the elaborations, the self-deceptions, distractions, and remove the hindrances from the Being-ness of the awakened mind interconnected with everything on the Earth. Then he cultivates enlightenment factors such as mindfulness, examination of his states of mind, tranquility, vitality, concentration, equanimity. Finally there are no separate perceptions. Clearly, his mind and feelings are alien to the so-called civilized and modern people.

In the emptiness of mind, feelings and reality remaining after severe purification, there are levels of reality which are not detectable by ordinary untaught people. There are levels of mind, hinted at in

Appendix D: An Attempt to Label Some of the Separate Levels of Mind and Consciousness.

These levels of mind become the monk's ordinary reality. His reality is not detectable to the untaught mind that is unconditioned in Buddhist purification. These levels of mind operate on several mental archetypes such as curiosity and investigation so that many hypotheses and assumptions arise that rarely occur to the untaught, civilized and modern people, drowning in the senses and the deception of thinking. They are unconditioned in Buddhist purification. Read Appendix A: Assumptions and Hypotheses are Introduced to Reduce the Problem of Intercepting the Communications from the Underlying Reality and the Problem of Decoding the Information Received.

After you have read Appendix A, you will realize that the ideas in this book rarely occur to those who have not made a sustained effort to remove the layers of elaboration. These ideas of seeking the information within the communications from underlying reality to the earth are alien from those

> who are addicted to being in a hurry and then laying about in sloth, addicted to feelings caused by drugs, and
> addicted to enjoyment of hating and
> participating in the mania demanded by propaganda that advertise products and
> thrilling in the government demands for men to die in war, and feeling superior by wearing special clothes

The ordinary untaught people could be labeled, "automata." They have neither interest in such levels of mind nor the investigations that arise in the purified mind.

I suggest that you accept the goal of discovering the vast region of the human mind. Then, you will perceive the realities hinted at within this book in a different way.

I suggest that you investigate the following hypotheses.

Hypothesis: The underlying reality communicates the laws and the properties of the universe.

Hypothesis: The underlying reality communicates to the whole of the Earth. The entire Earth and all its living beings are meant to continue; not just humans.

Hypothesis: The underlying reality communicates that humans are one type of living being; a part of the whole Earth and humans are responsible for making efforts to help the Earth continue.

This book arose in a higher level of mind that is available to each person who removes the layers of elaborations that are so-called civilized, modern life. Such persons can then purify their minds to remove the hindrances and to begin the mental habits of enlightenment.

Human archetypes exist. These are tendencies in most human minds that drive most people to behave in certain patterns. Virtually all human speech and actions are initiated in the mind. Some archetypes manifest as hierarchy, drive for power, war, construction of community property, including systems that preserve the continuation of the community. An archetype results in various behaviors that result in destruction of community property along with the interconnecting systems.

A tiny part of human kind, the chiefs at the top of the hierarchy holds virtually all power. The tiny part continues to lust for power. They are the chiefs of the corporations, the governments and the religions that persuade the ordinary untaught people to be civilized and modern. This drive for power archetype is destroying an enormous number of communities, is destroying the Earth and threatens to kill all living beings.

All human kind must invent a way of living that manifests in the chiefs at the top hierarchy as power to continue the Earth, as cooperation, and negotiation instead of destruction of the Earth, violence and war. This is the reality which exists now.

Hypothesis: The underlying reality is communicating that we must recognize the threat that humans cause the Earth.

This book provides the guidance and the major methods to attain two goals.

One research goal is profits from knowledge of the communication with the underlying reality. The intent is to discover the total reception of, decoding of, and correct use of all communications from the underlying reality.

The sum of all the suggestions mentioned herein for locating, intercepting, observing, and decoding communications from the underlying reality, are not the only efforts to communicate. These efforts will yield the valuable information contained in the communications. After gaining information from this communication, one must wisely use the information gained. The value of these endeavors is in the profits gained from using the information.

The guidance and methods provided herein are only a part of all possible channels and communications from the underlying reality.

The other research goal is knowledge of all the capabilities of the human mind, especially extra sensory perception (ESP). The intent is to discover the total powers of and the correct use of all the powers of liberated minds.

The justification for the publication of this book is to persuade groups of people to pursue the recommended goals, to accept the conclusions, and to use the information from the underlying reality for extending all life on Earth by attenuating the human archetypes that manifest as lust for power and the intent to destroy.

Recommendations

Massive resources that have been wasted on extending the power and prestige of a tiny number of elite people it government, business organizations, and religion for the last 3000 years. The major tool used to project this power and prestige has been killing and destroying the Earth with weapons and war. The many tools for expanding the mentally ill egos of the elites are now threatening the existence of life on Earth. This intention to amplify the egos of a few elite people must be diverted into the intention to discover what will aid the evolution of human kind in concert with all life on Earth. The guidance toward discovery is as follows. Think, speak, and act to promote the following.

Whatever heals the Earth and whatever extends all life on earth.
Whatever brings people together under the principles of Respect, Harmony, Tranquility, and Purity.
Whatever aids the evolution of human kind, in concert with all life on Earth, toward the following.
Correct view of reality,

Correct thinking,
Correct speech,
Correct, action,
Correct mindfulness,
Correct concentration,
Correct equanimity,
Correct liberation from suffering,
Correct pleasures,
Discovery of, total reception of, decoding of, and correct use of all communications from the underlying reality,
Discovery of the total powers of and the correct use of all the powers of liberated minds.

I recommend the diversion of all money and human effort now wasted in weapons, war, threats of war, threats of destruction of parts of the Earth held by large groups of people, and wasted on weapons and armed forces of destruction of the means of production held by large groups of people to promote the above guidance. I recommend this money and effort be directed toward valuable goals. The guidance is listed above for attaining the goals.

Introduction

How to Intercept and Decode the Communications from the Many Levels of Underlying Reality

Considering the many thousands of years and the number of efforts that have been put forth trying to detect the underlying reality and to define its attributes, there has been meager success.

Historical methods of describing the information communication from the underlying reality to the environment are usually based on convenient assumptions that cannot be tested.

Methods are presented herein to intercept the communications from the underlying reality. Whatever is discovered in this reduced set of research will be a part of the entire collection of all communications discovered from the underlying reality.

One could assume that discovered patterns are the decoded communication signals transmitted into the fragment of reality that is within the limited realization of humans. There is a fundamental limit on human discovery of the underlying reality. Religions and science have sought to discover and describe the underlying reality. The method of gaining subtle and gross knowledge is explained.

Appendix A Assumptions and Hypotheses are Introduced to Reduce the Problem of Intercepting the Communications from the Underlying Reality and the Problem of Decoding the Information Received

This is focused on limiting the investigation so it is possible to make progress. The approach to research is to consider a subset of all existing communications to the world from the underlying reality.

Some of the taxonomy of investigations into the communications from the underlying reality is defined. Hypotheses and assumptions are presented to reduce the problem of intercepting the underlying reality. There are many layers of underlying reality. There are many channels through which the communications may take place. By proposing assumptions and hypotheses and testing them, the information being communicated will be decoded.

Appendix B Some Methods for Removing Noise from the Communications from the Underlying Reality

How to Interpret and Decode the Communications from the Many Levels of Underlying Reality

Perhaps, the other mechanism of interpreting the underlying reality. Realizing a physical individual reality, there is another mechanism which the communication may take place. By decoding beyond biochemical and electromagnetic information to a consciousness level of the brain.

Perhaps, you decode what this is. Removing from individual the communication from the Underlying Reality.

This is focused on one aspect of discovering the underlying reality. This research is further limited to methods of removing noise. This aspect is removing noise from the communications. Discovering the underlying reality is vast and complex thus it must be simplified. This article is mainly about communications that have a wave form. These are easier to investigate due to existing electronic tools. The approach is limited to a search of wave based communications with amplitude modulation.

To achieve the decoding of the intercepted communications, the noise must be removed.

The following concepts are for the purpose of removing noise from the information within a communication from the underlying reality. They are based on methods and equipment used to remove noise from radio, television, and other electromagnetic (EM) communications. The methods and equipment presented are restricted to the EM wave forms. The knowledge of how to decode wave borne information can be extended by mathematics which is a language of analogy.

The intent of presenting the methods, concepts and mathematics used in EM signal processing is to stimulate the imagination to use these as analogies for other channels through which communications occur. For example, the properties of the underlying reality may be directly transmitted into the world through changes in species, changes in molecules, changes in human abilities to perceive, to invent new ideas or to produce animals with more functions. There may be a communication that yields less function due to noise in the signal. There may be transient noise in the communications that result in reproduction that temporarily produces a few mutants with a poor survival rate or no ability to reproduce.

The scientific explorations into the underlying reality are briefly noted. Is mathematics inherent in the underlying reality? Consider how research into quantum physics may increase the information available for experimenting with methods of intercepting the underlying reality. There is strong pair creation in strong fields.

Can the quantum theory of wave fluctuations and uncertainty be extended to include creating psychic particles with no mass? An hypothesis and an experimental design for testing it are proposed.

One could study matter in the void, ether, geometric theories and the dimensions of space and time, the metric field, and the vacuum complexity. Is the matter in the void: ether, space, and mind field the underlying reality? Are space, time, the metric field, and the vacuum complexity the underlying reality or is there another reality even more fundamental? Those who have scrutinized quantum mechanics realize the limits to knowing the underlying reality. Geometric theories and the dimensions of space and time are more fundamental than most explanations of the underlying reality. What is the underlying reality of the universe from the point of view of some physicists?

There are few instruments to aid the limited investigative abilities of humans. Many layers of instruments and mathematics were invented to overcome the limitations of human senses. Architecture and Feng Shui, for example, increase in precision in proportion to the accuracy of the decoding of underlying reality. Diffusion is another possible channel of communication. Decoding the many channels of communication may succeed by using a higher level code. The scientific answer to these questions can be found.

The scientific search for the underlying reality was based in the material world while the Buddhist search was based in pure mind. Buddhist influences also influenced the research into the underlying reality. Consider that Lao Tzu decoded the Tao using Taoist principles.

Fundamental tendencies of all societies are to obstruct adaptations of theories or changes in paradigm

Examples of failures to decode the communications of the underlying reality are presented. A method of decoding the underlying reality is to deny its existence. An experiment is proposed to verify or falsify existing mental constructions describing the underlying reality of the universe.

A problem of communication is to accurately determine the bare information transmitted in the messages, by identifying the noise and then subtracting it from the total message. Several recommended mathematical approaches are presented for decoding the reduced set of communications from the underlying reality. Clues to decoding communications from the underlying reality are suggested by advanced digital signal processing and noise reduction.

Examples from outside the scientific approach are presented. Analogies are presented to aid the perception of the problem from several angles.

Appendix C: Many Different Varieties of Reality.

This is a presentation of many different views of what constitutes reality. The reality of the ancient Aztecs, the reality of Jewish people who entered the NAZI concentration camps as well as well as the scientific structure of reality, and virtual reality are mentioned. A list of books describing many realities is offered.

Appendix D: An Attempt to Label Some of the Separate Levels of Mind and Consciousness.

This appendix suggests adequate mental tools for locating where and when communication with the underlying reality is found.

Appendix E. What is Chi?

This appendix notes early Chinese concepts of Chi.

Appendix F: Decoding Several Levels of Reality in the Moral Mazes of Thinking, Speaking, and Acting by Chiefs in Governments, Corporations, and Religions

Appendix F exposes the levels of reality which the chiefs of organizations must operate within.

Next, it proposes to divert resources away from war and the destruction of the earth and to direct them toward ethical use of human intelligence, toward studying the human mind and psyche, and toward Buddhist concentration, meditation and absorption. Researches using these resources are proposed.

One research goal is knowledge of the communication with the underlying reality. The intent is to discover the total reception of, decoding of, and correct use of all communications from the underlying reality.

The other research goal is knowledge of all the capabilities of the human mind, especially extra sensory perception. The intent is to discover the total powers of and the correct use of all the powers of liberated minds.

The main objective of this book is to be the engine of directing mental activity, funding and concentration on the power of the mind, on consciousness based on the human body, and on knowledge of the underlying reality.

The explanation of the title is found in Appendix G. Li Bai lived in several realities, the Emperor's retinue, alcoholic drunkeness, poetic musings, the Emperor's consort Yang Guifei who was an immortal beauty disguised as a commoner, the lake appearing as a solid walkway, and the moon disguised as his lover Yang. Another level of reality is the presentation of the story as the conversation between two lovers; the woman is playing the role of Yang Guifei and the man is playing the role of Li Bai.

Chapter 1

Attempts by religious organizations and the early Greek philosophers to define the underlying reality

Religions and science have sought to discover and describe the underlying reality

Most religious systems include a search for the underlying reality, God, Tao, Jehovah, the Great Spirit, etc. Each has many traditional concepts about the properties and the communications from the underlying reality to humans. The Christian and Jewish religions postulate a God, Jehovah that existed before the universe. God communicated the universe into existence and communicated the laws and properties of the universe. The ancient Greeks and the ancient Indian religions postulated several gods as underlying realities which caused the existence of the world, caused all the changes in people and other living things, ordered the changes for inanimate objects, and set out laws for living and inanimate things. These gods had distinct personalities that were drawn from human examples. The gods interacted with each other.

Organized religions invented creation stories and sought to know the underlying reality God, Tao or other names

There is an archetype of the human mind that is expressed as curiosity to know more. Some members of religious organizations wanted to know elementary causes of their observations of the world. There was a desire to manipulate the elementary causes to improve the living conditions of humans. Some men identified that the observations of the world were delivered to the mind by the senses. There was a pattern of curiosity about the world in combination with the willpower to manipulate the elementary causes to improve the environment. This curiosity and willpower stimulated the origins of many arts and sciences.

There are unknown influences on weather, events, accidents, and human fate. The desire to control the weather and other events was a driving power to understand underlying factors which were not capable

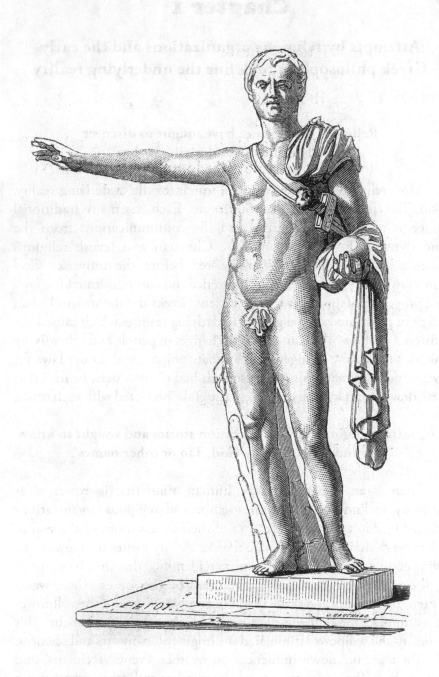
J. PETOT.   J. GAUCHARD.

of being sensed. Some men assumed that there was a spirit that caused weather. They also assumed that the spirit used weather to punish or reward humans depending on human thinking, speech and physical actions. They assumed there was the potential to know the mechanism of weather.

The transformation of potential into wisdom about the spirit required men to decode the intentions of spirit. If one could know how to decode the communications of the underlying reality, then one could communicate back to the underlying reality, thus controlling the weather. These were the preoccupations of those who loved knowledge. Some people wanted power over the weather and other things such as control over powerful men. This process of loving knowledge, philosophy, is generalized in much of this book.

Hypothesis: the archetypes in the human non-conscious diffuse through consciousness into the mind.

For a detailed description of the diffusion within the psyche including the mathematics see Howard, Robert G. (2012c).

It is possible that awareness in the mind of the archetype was equated to communion with God, Tao, spirit or whatever was the local concept of the underlying reality. This could be called the archetype of religion. Religion often uses analogy to explain complex concepts. Analogies are used often in this book as aids to the imagination.

The analogy of the engine

The potential to know the encoded message from the underlying reality is analogous to an engine that causes movement. The curiosity to know is analogous to the trigger that starts the engine. The willpower and curiosity diffuses into all parts of the body, consciousness, and mind of the actor. The potential within the engine is transmuted into willpower which is expressed in thoughts, speech and bodily actions which, collected together, result in events. These are the factors that create Popper's World3; mental objects that exist only in the mind. (Popper and Eccles, 1977) Feng Shui is a World3 engine that evolved from this trigger and was nourished by this love of knowledge.

An element of World3 is the desire to seek and identify underlying reality. In addition to religious organizations seeking, there are science organizations seeking the origin of all physical things and all living things. Out of science comes technology also seeking, for example, how can a metal be made stronger, how can a food crop grow faster, or how can things be made to move faster. What are the underlying properties that answer these questions?

Assume there is an underlying source of reality. The search for an underlying source of reality has been conducted for thousands of years within a religious context.

The Jews in Judea in 900 BC had a defined concept of Jehovah. The origin of this concept was not stated. The earliest documented search was by Lao Tzu in about 600BC. (Kohn and LaFargue, 1998) The question remains, "What is Tao? Watts commented on this question. (Watts, 2000) Lederman asked a slightly different question in a scientific context, *The God Particle: If the Universe is the Answer, What is the Question?* (Lederman, 2006) A recent attempt to answer a similar question was conducted as a scientific effort, *Decoding Reality: the Universe as Quantum Information.* (Vedral, 2010)

One can conclude that the results of these searches are vague due to this world which is complicated and impossible for the human mind to understand in depth. Scientists investigate the world to discover parts of the underlying source of reality. Scientific research explores the easier questions about what is reality and how much one can know about reality. But there are always accidents that do not make sense. One could say the scientists explore accidents which make sense under the pressure of curiosity, thus simple laws can be defined. There are always complications, even chaos, so scientists abstract a reduced part out of the whole reality; a non-accidental domain in which simple laws are found. The complications are called initial conditions and the domains of regularities, not accidents, are called laws of nature. Such a description of reality has limits but it allows the natural sciences to discover some of the laws. (Wigner, 1967)

Religious organizations were the primary receiver of causal connections from the channel of information from the gods to the earth and to man. Religion has interpreted the information in terms of effects on earth and humans. Religion organized the search for communication with God and Tao over centuries of years and over many members of the religion.

Ancient Greeks asked the question,
"What is the ultimate composition of reality?"

The question of what the world was composed was asked long ago. Thales of Miletus (c. 624 – c. 546 BC) attempted to explain natural phenomena without reference to mythology. He provided an explanation of ultimate substance, change, and the existence of the world. This became an essential idea for the scientific method. He used mathematics as a tool to describe observations. Democritus (c. 460 – c. 370 BC) was a Greek philosopher born in Abdera, Thrace, in what is now Greece. He was accredited with constructing a basic underlying reality composed of atoms that are physically indivisible; that between atoms lies empty space; that atoms are indestructible; that they have always been, and always will be, in motion.

We can conclude that all of existing reality known to humans is "information" which we can label, "shadows of the originating information." This refers to Plato's "Allegory of the Cave" in *The Republic*. (Plato, 1959) Plato illustrated that humans are severely limited in their meager sense data of the environment outside their bodies.

There exist many explanations of the source of reality which cannot be detected by human senses. Spiritual leaders in almost all societies have invented a dogma about the underlying source of reality.

The analogy of the farmer

Ancient Hindu sages, such as Patanjali, proposed an analogy of the inherent unfolding of nature within all existing entities especially humans.

The transformation of one species into another is caused by the inflowing of nature. Good or bad deeds are not the direct causes of the transformation. They only act as breakers of the obstacles to natural evolution. Just as a farmer breaks down the obstacles in a water course so the water flows through by its own nature. The farmer who irrigates his field from a reservoir does not have to carry the water. The water moves by itself. The farmer only has to open a gate or break down a dam and the water flows by the natural force of gravity. In this analogy, water moving by gravity is the force of evolution which each living entity carries within itself waiting to be released from the reservoir. A person's acts open the gate. The water runs into the field. The field bears its crop. It is thereby transformed. The form of the next generation, or re-incarnation, is determined by the nature contained within the human species. All progress and power is in each person. Perfection is in every person's nature. Only it is dammed in and prevented from taking its proper course. When a person can open the dam in the correct place, nature rushes in.

The performance of bad deeds and the consequent accumulation of bad karma break the dam around the reservoir in a place that results in a disastrous flood which ruins the growing field. The water flows according to its response to gravity. The water has to be properly directed by the farmer.

The true underlying reality of evolution is the manifestation of the perfection which is already contained within each living being. In most cases, the perfection is dammed and the infinite reservoir is struggling to express itself. Darwin's theory of survival of the fittest is not necessary for one person to progress toward perfection. In each human, there is a potential higher level of existence. (Prabhavananda and Isherwood, 1953, p. 203-206)

Chapter 2

Attempts by scientific professions to decode the underlying reality

Scientific search for the source of mass, called the Higgs Boson

An objective of scientific endeavors is to uncover the chain of cause and effect. When the causal connection is in doubt, probability mathematics often supplies a handle on the influences on an event. The scientific approach formally accepts layers of underlying reality or any other uncertain sources of reality. See Appendix C: Many Different Varieties of Reality.

Questions about the underlying reality directed toward scientific efforts

Is there a non-physical source, which is labeled underlying reality, that encodes some fundamental tendencies and information and then transmits them as laws and characteristics of the physical and mental worlds?

Does the underlying reality transmit laws of physics through several communication channels into the physical world? Or is the underlying reality more fundamental than the physical and biological laws?

Does the underlying reality enable the tendencies and information it transmits into the universe to be decoded by the physical receiver?

Do the decoded information and the tendencies result in order, processes, chaos, and disorder in the physical universe?

Aspects of the scientific approach to discovering the underlying reality

Laws of chemical combinations have been framed. But no transformation of a combination of non-living elements into living beings, consciousness, or mind has been accepted. Most scientists stand in awe of the complexity of living entities and life. They perceive beauty

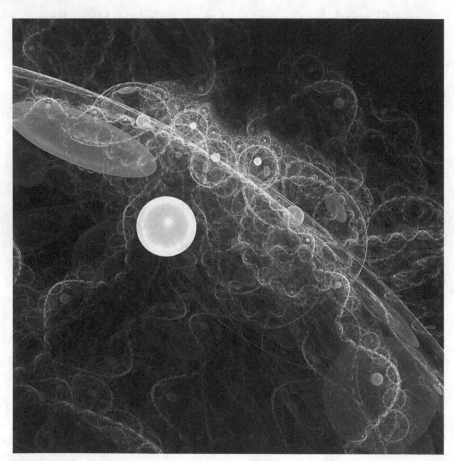

in the discoveries of science. They are loyal to the dogmas of science. They experience strong feelings about the achievements of the sciences. This is the way the contemporaries of Thales, the Greek, felt about the myths of world creation thousands of years ago. These feelings by those in the science industry and its results have persisted for thousands of years.

Assume that the fundamental law of all change is a decoded principle of underlying reality

One can conclude that all things change. Is change in itself an instruction from the underlying reality? Perhaps change is an encoded communication from underlying reality that is decoded by all things. Change is embedded in all things. Change emerges in all things because all things decode the communication from the underlying reality.

Another fundamental law is diffusion. Humans are aware of diffusion of heat due to observing things heating up. Humans are aware of mass diffusion in rivers and the crumbling of mountains. Language and knowledge diffuse into different territories. All discretely different things near each other diffuse into each other and change as time increases. This is also true for human mental hypotheses, beliefs, and mental abilities. The mathematics describing diffusion is based on the Gauss Divergence Theorem. Denbigh provided the details of several types of diffusion embedded in wave phenomena. (Denbigh, 1951)

The Gauss Divergence Theorem is a tool for describing diffusion. It is usually stated mathematically, but it is stated below in natural language as applied to diffusion in a human mind-consciousness-body system in which knowledge, K, is diffusing. The theorem states that knowledge stored in the system is the amount inherited at birth plus the amount that has diffused into the system minus the amount that has diffused out of the system plus the knowledge internally generated. Howard provided the details of the mathematics. (Howard, 2012c). Popper and Eccles addressed the fact of being born with knowledge or skills already part of the mind-brain. (Popper and Eccles, 1983, p. 120-124)

This diffusion was not always obvious. Many years and large numbers of investigators decoded some layers of the underlying reality. This resulted in the conclusion that the universal phenomenon of diffusion underlies the entire environment including living entities. Mikhailov and Ozisik cataloged the detailed overview of the extensive mathematics of all heat, momentum and mass diffusion. This math can be applied to other diffusion processes such as electrons in electricity, and sense data in the nervous system. The principle is valid for diffusion of ideas in the mind. (Mikhailov, M. D. and Ozisik, M. N., 1984)

By decoding the communication from the underlying reality, which was an encoded message, diffusion was identified. What can we discover about an underlying reality that manifests as diffusion? Is there a principle or property of underlying reality, represented by diffusion, that was encoded and communicated into the world but has not yet been identified? Then everything decodes the communication which established universal diffusion and obeys the requirement for diffusion.

Quantum theory was so mysterious that many physicists decided to embody a fundamental conceptual structure for use in interpreting research. The laws of quantum reality were believed by some physicists to be the underlying reality.

One of the principles is the uncertainty principle which observes that there are complementary properties at the quantum level such that only one of them can be known accurately because the measuring process disturbs what is being measured.

Another principle is that subatomic particles do not exist as determined entities but as a state function embodied in probability mathematics. The astonishing feature of the state function is that the particle exists in many states simultaneously and also exists in none of the states. Also, the thinking of the experimenter and the theoretician yield the temptation to read the experiments and the written theories in accord with their prejudiced thinking. So thinking affects the observations.

The main interpretation of experimental and theoretical data was formulated as follows. The Copenhagen Interpretation from about 100 years ago is similar to the Tao Te Ching from ancient China. (Kohn and LaFargue, 1998) (Kosso, 1998)

1. Quantum mechanics is complete. Indeterminate entities and superposition of many states of existence in a single location (the state function of probability of an entity existing in a set of probable superimposed states) are facts of the quantum physicists but they cannot be tested or falsified because they are not observable.

2. Any observable entities appear in the large scale classical theory.
3. The task of science it to describe relations between observable entities. One can only speculate about unobservable entities but that is not science.
4. One cannot ascribe any reality to the quantum state function or the quantum world.

Number 1 goes back to the religion approach of formulating an hypothesis about Nature or underlying reality that cannot be observed and insist that it is true and not questionable. This is superstition and yields ignorance.

The future behavior of living environments is almost impossible to predict.

The discoveries of quantum physics remind one to allow for probable influences; not to insist on exactly determined answers to questions about the underlying reality. The mathematics of probability has been expanded for hundreds of years. Some of the math is based on large numbers of exactly defined sets of repeatable experiments. In addition there are experimental probabilities based on observations of experiments. However, a situation in which a human finds himself is almost always a single event or experiment that is not clearly defined. The influences on the event are unknown.

These memories suggest caution in identifying the channel of communication from the underlying reality. Realizing these factors in the search for the channel, one can use probability math and quantum physics as mental tools to identify the channel. On the other hand, the probability of an event, lacking the certainty of an event, and the unknown influences on an event are additional factors which obscure the information from the underlying reality and confuse the channel. Refer to the discussion of the probable and unknown influences treated by the *I Ching* in Chapter 8.

Another interpretation from quantum physics yields the hypothesis that nothing exists for humans until they observe it and measure it. This is a serious consideration in spite of its apparent nonsense.

Can one infer the underlying reality from observations and appearances? This interpretation is still being debated. Another way of stating this could be, "Humans have sense experiences which include reading measurement instruments. The sense experience and the measurement supplied by the instrument are encoded expressions of the underlying reality. Is it possible to discover the attributes of the underlying reality by analyzing all the encoded signals received from the underlying reality? Are there non-encoded communications from the underlying reality? Can the expressions be decoded to reveal the properties of the underlying reality?"

There are several interpretations of the quantum theory and its experimental evidence. New interpretations are commonly proposed and debated. Niels Bohr, the leader of the quantum theory researchers, described possible objectives of physics in general.

1. One could describe Nature as it really is. [This is labeled 'decoding the underlying reality' herein.]
2. One could describe what one knows about Nature [which could be believed to be the encoded underlying reality.]
3. One could describe how humans participate with Nature; how humans conceive of Nature; how humans take action in physical Nature. [These are more directly known by experience] (Kosso, 1998, p. 29)

Bohr also wrote, "There is no quantum world; only an abstract description. It is wrong to think the task of physics is to find out how Nature is. Physics concerns what we can say about Nature."

Another contrasting definition of physics was proposed by Einstein, "Physics is an attempt conceptually to grasp reality as it is thought; independently of its being conceived. In this sense, one speaks of physical reality." (Kosso, 1998, p. 177)

Refer to the *Tao Te Ching* by Lao Tzu for comparison of the similarities. (Kohn and LaFargue, 1998)

The physics of the mental construct of empty space

Consider empty space with no mass but containing electromagnetic radiation, fields, energy, and other factors. Consider that it may generate energy, particles, space and time. Could it also be a channel for communications from the underlying reality?

Sir Isaac Newton laid the groundwork for modern natural science in 1687. Before him, no one had tried to differentiate between the realm of initial conditions over which we have some control, and that of the laws of nature over which we have no control. After the initial conditions are set for a rock, the laws will fix the rock's trajectory.

We can characterize the curvature of empty space with angles in a triangle. When the sum of the three angles in a triangle equals 180 then the curvature is zero and the space is flat. The theory of relativity refers to the metric field with a curvature in four dimensions, three lengths and one time. A four dimensional space-time may contain fields such as a gravitational field. So it is not empty. Einstein's point of view was that the field defines the space and the masses in the space define the curvature. So space is a properly modified expression for the positional quality of the universe of masses. If there are no masses, then there is no space. Gauss invented the "Gauss metric" to describe this frame of reference. (Genz, 1999, p. 173-178)

Even in a vacuum, there is the complexity of quantum theory. Blackbody radiation at zero temperature in a vacuum has been understood to yield the Casimir effect. If there are fluctuations in a vacuum that sometimes yield real particles with momentum, then the will be an equal and opposite reaction to the momentum. This could be measured.

In addition, the math reveals that the "blackbody" at zero degrees K has infinite energy. This was also predicted in Einstein's general theory of relativity. (Genz, 1999, p. 189) In this state, a sub-atomic particle with completely known velocity has completely unknown energy, E defined as,

$$E = mc^2 + \text{potential energy} + \text{kinetic energy}$$

All of which are fluctuating. The fluctuations can be observed in the Fourier transform of light waves. *Fourier Series and Optical Transform Techniques in Contemporary Optics.* (Wilson, R. G., 1995).

This was the conclusion that Lao Tzu held also. (Kohn and LaFargue, 1998)

There is an important fact with many ramifications: light scatters light. The fluctuating charges in the vacuum are also the reason that light rays cannot simply penetrate each other without perturbing each other. And this is so for all wavelengths of electromagnetic radiation.

Photons, a range of wavelengths of electromagnetic radiation, are conjectured to couple to the electric charge. Electric charges emit and absorb light. All light effects are conjectured to be reduced to this elementary process. This had not been experimentally proven or falsified as of 1999. The scattering of light in the strong electric field of an atomic nucleus was experimentally verified with gamma rays of very short wavelengths. It is conjectured that virtual particle-anti-particle pairs fluctuating in the vacuum also permit the fusion of the two photons into a single real photon in the presence of an external field. (Genz, 1999, p. 247-248)

If a high energy particle and an equal but opposite anti-particle are generated by a fluctuation, they have a very short existence before they are canceled by each other. If the particle has a low energy, it may not be canceled for a million years. This could also be extrapolated to yield any atom of an element. Where are the anti-particles lurking in wait to cancel out the particles? However, this has neither been verified nor falsified experimentally. (Genz, 1999, p.249)

Consider spontaneous creation of an electron –positron pair in a strong electric field generated by a strong positive electric charge in a vacuum. This is spontaneous creation due to the fluctuations. The term 'virtual' here means that these particles vanish immediately upon appearance as explained above.

The charge surrounded by virtual electrons and positrons. Then, physicists conjecture that a fluctuation in a vacuum can generate

a virtual electron + energy→real electron
and virtual positron + energy→real positron

The energy of the type $E=mc^2$ is contemplated in this combination reaction. This energy, E corresponds to the relativistic relationship between mass and energy. The relationship requires that the mass, m, of an object increases when a mass is accelerated to approximately the speed of light, c. When this energy is added to the virtual electron then a real electron created. When this energy is added to the virtual positron then a real electron is created.

The real electron then is conjectured to join with the real positron to yield a hydrogen atom. When this energy is added to the virtual positron then a real positron is created.

The next reaction is

real electron + positron →hydrogen atom.

An atom of hydrogen is created out of a fluctuation and the energy from a nearby fast mass, m. This could also be extrapolated to yield any atom of a heavier element. This has neither been verified nor falsified experimentally.

This is a simple demonstration of matter being created out of a void.

Quantum theory is not always confirmed by common sense experience. Much of physics is formulated in the language of mathematics. This obscures the concepts and the mental process of inventing the physics. At times, the topics of physics imply that the common sense world is not real or that the physicists are not living in the real world. For example, the uncertainty principle in quantum mechanics and also the theory of relativity lead to a hypothetical relationship of time and length. This yields a problem with decoding the communications from the underlying reality. The ability to decode the information requires a large taxonomy of ideas and the willingness to accept what is not in harmony with scientific dogma.

There are strange ideas about breaking the symmetry of the vacuum. The ideas are named after the people who formulated them, Jeffrey Goldstone and Peter Higgs. To understand them, one must first imagine the objects of interest and increase one's vocabulary. This allows one to go beyond past experiences and to expand the limits of one's imagination. Many of the mind objects in quantum physics are not in one's experience, such as particles with no measured position. But they are presented as analogs to our experience. (Genz, 1999, p. 257)

The general relativity theory has a geometric formulation of the gravitational force. It serves as an example for other theories of interactions. The concept is that the trajectories of masses are the shortest path between two points in four dimensions. This is an insight into the communication from the void of underlying reality.

Kaluza and Klein expanded the theory of relativity to include the electromagnetic interactions. They introduced a second dimension of time which can be called, 'the long view.' History embodies the long view of time, t5, which is all the changes and history of one event perceived simultaneously in one frame of view. This introduces the concept that there is another dimension of time, t5 that most humans do not experience. However, anyone can experience two dimensions of time with a little training.

There are many properties that humans cannot experience but we can try to decode the evidence we are capable of experiencing. These ideas are outside the realm of common sense. If we allow facts to exist outside of common sense and sense experience, then we can infer some of the underlying reality. This could be the goal of science and quantum mechanics in particular. (Pais, 1982, p. 329-336)

The Chinese art and science of Feng Shui, similar to the American art and science of architecture, depends on an accurate decoding of the details of space and time

Consider an example of professions that require detailed information on the world including past, current and future influences: the design of a building or landscape is based on detailed analysis. The site will exist for centuries. Thus, great detail must be decoded from the past, future and the conditions at the site. Only an advanced and specialized group within a complex society could devise Feng Shui, architecture, and the many other arts and sciences. Let us speculate on how an established system of decoding the sense data from the world could yield such arts and sciences.

Since creation stories are common in all groups of humans. It would be reasonable to speculate that the Chinese asked where the influences in the taxonomy of Feng Shui came from. They may have asked questions, "What does Heaven invent in the Earth and Man?" We could translate this as, "What is the description of the information communicated from the underlying reality to the environment surrounding us?" The answer would have been based on minimum assumptions.

Wave components of underlying reality were commonly observed. They were even represented in ceramic art thousands of years ago. Since waves were so common, a minimum assumption is that wave phenomena were part of the decoded communications from the underlying reality. When the ancients invented Feng Shui as an insight into the influences of underlying reality, wave motion may have been included. In fact, Feng Shui translates as Wind Water. Then the underlying reality would have wave attributes. All the waves and all the inherent fluctuating character of most things were conjectured to be interconnected.

The preponderance of wave phenomena has been investigated extensively. Coulson demonstrated the application of wave math to many observations. (Coulson, 1977)

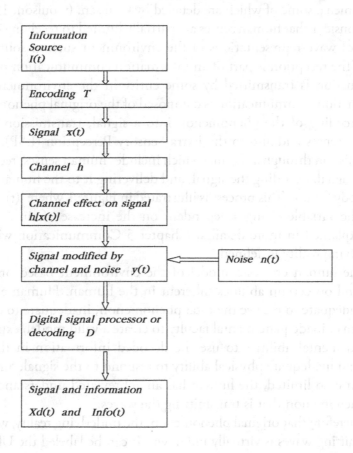

Figure 1 *Generalized Communication and Signal Processing System*

(t) means the function depends on time, t

Waves are a channel for the many communications transmitted by the underlying reality

Many communications have the form of a wave or are associated with a wave component. This book mainly considers communications from the underlying reality, the total environment, which meet this requirement; association with a wave. Wave mathematics describes a diverse set of phenomena, some of which are detailed by Coulsen. (Coulson, 1977)

Consider that human senses are partially limited to receiving certain types of wave representations of the environment such as sound and light. The reception is part of an information communication process. Information is transmitted by some entity in the environment. The information communication is composed of the original phenomenon, the encoding of the phenomenon into a signal, transmission to the human senses and also to the Extra Sensory Perception (ESP), noise, transmission through a channel which includes human senses, reception of the signal, decoding the signal, and delivering it to the human mind and body for use. This process is illustrated in Figure 1 where '(t)' means that the variable changes dependent on the increase of time. Figure 1 is explained in more detail at Chapter 5 Communication with the underlying reality is defined.

The human creates a model of the environment based on sense data and on certain abilities inherent in the human. Human abilities are inadequate to receive the total phenomenon, inadequate to decode the signal, inadequate mental faculty to create a valid model, extremely limited mental abilities to use the decoded information in the best way, and inadequate physical ability to respond to the signal. Since the human is so limited, the human has an abbreviated understanding of the phenomenon that is transmitting the waves.

Therefore that original phenomenon, the underlying reality, which is transmitting waves is virtually unknown. It can be labeled the Ultimate Source of Reality. It is also labeled God, Brahman, Tao and so on. Some of these labels are given attributes. The Tao is not given many attributes. In this book, most of the attributes assigned to the unknown source of reality phenomena in the environment are excluded. Consider what could be known about the label with the least attributes; underlying reality.

An example of a phenomenon with a recurring property is the wave. (Coulson, 1977). Any fluctuating phenomena can be described mathematically with wave equations. One method of manipulating an infinite number of waves simultaneously is Fourier analysis which has become a valuable tool of research. Fourier analysis is being embodied in electronic analysis tools so research can ponder very complex wave phenomena in the environment. (Pinsky, 2008) (Wilson, 1995) One could say that wave components of underlying reality are commonly observed. Since waves are so common, one could propose:

Hypothesis: waves are part of the decoded communications from the environment.

Unless falsified, aspects of the underlying reality have wave attributes. All the waves and all the inherent fluctuating character of most things are interconnected. Wave phenomena influence other wave phenomena

Hypothesis: a waveform is a subset of all possible channels through which the underlying reality transmits communications.

Some of the artisans and elite dignitaries of the Chinese contemporary to the founding of the Chinese Qin Empire in 221 BC recognized many types of wave motions. Europeans and Americans have investigated wave motions mathematically in great depth for hundreds of years. Communications and other technologies are based on knowledge of waves. The mathematics of Fourier series and wavelets were invented to represent collections of wave motions. Many separate patterns of parameters can be joined together mathematically to explore the collections of wave motions. See

Introduction to Fourier Analysis and Wavelets, (Pinsky, 2008)

Extraction of Signals from Noise, (Vaĭnshteĭn, 1962*).*

Advanced Digital Signal Processing and Noise Reduction, (Vaseghi, 2006).

Communications transmitted by the underlying reality

A mental function of the architect or Feng Shui practitioner is the ability to understand the underlying reality in a house, building, or landscape. The best practitioner would have some training in gathering the information from his senses and also from the underlying reality. His observations would be processed to decode the intentions of the underlying reality or the approximate tendencies of the underlying reality. In Chinese culture, chaos is the opposite of the ideal condition, stability. So a Chinese Feng Shui master may be biased to interpret stability whereas an American architect may be biased toward perceiving changes; welcoming chaos. Obviously, predicting future stability or chaos is difficult or impossible.

Part of the manifestation of underlying reality, can be described as the entire communication process itself. Long after the arts and sciences of architecture and Feng Shui were formulated, the mathematics of communication was described in terms of information theory. (Shannon, 1948) Thus, information theory is another tool for discovering the underlying reality and its attributes.

Using waves to reduce the problem of decoding communications

One can reduce the difficulty of decoding the communications from reality. One can conceive of the communications from the underlying reality to be information only and leave the actual changes of earth and man out of consideration. Other methods can be used to reduce the difficulty. This method of reducing the entire interrelated system yields an insight into part of the manifestation of the underlying reality.

Another method is to remove all noise from the communication See Appendix B: Some Methods for Removing Noise from the Communications Received from the Underlying Reality

Hypothesis: limiting the set of the communications to those within certain assumptions and hypotheses does not distort the knowledge of all existing communications.

Outline of the mathematics of identifying and decoding the communication of information from the underlying reality

Communications technology is in the middle of an exponential era of discovery. It began with Maxwell's electromagnetic equations. (Maxwell, 1854) Almost all the communication inventions and systems are based on mathematics. It is impossible to explore communications technology without using the efficient language of mathematics. Shannon invented the mathematics and the definitions of the information concepts. The definitions of information and other terms will not be repeated herein. (Shannon, 1948, p. 379-423 and 623-656) Since the key to communications was published by Shannon, there has been an exponential growth of publications in great detail. Luenberger provided an introductory textbook. (Luenberger, 2006). There are new engineering disciplines which mostly explore the possibilities and apply the technology of communications.

However, Howard presented a short introduction to communications with a minimum of math. (Howard, 2012b, Chapter 11) A measure of information is the negative logarithm of a quantity which is a probability; essentially a negative entropy because it has all the properties that are associated with entropy. Processes that lose information are analogous to processes that gain entropy. This well developed mathematics can be used to refine the decoding of the communications from the underlying reality.

More mathematics for decoding the communication from the underlying reality

Time series mathematics is applied to slowly changing or extremely rapidly changing sequences of measurements. The problems to solve are recording, preservation of data, transmission, and use of information.

How is information measured? A simple case is the choice between two equally probable alternatives when one or the other is bound to happen. Another case is defining the information gained when one or more variables in the measuring system are fixed. Another case is when there is noise in the signal.

Let u = message from the underlying reality.

And let v = noise which could be generated by the mental habits of the observers: insistence on preconceived notions, misinterpretation of what the underlying reality is transmitting, poor observation skills, bad intent of the observer, incompetence in applying analysis equipment, and so on.

If v = 0 then u is infinite information from underlying reality.

When there is noise, information approaches zero rapidly as the noise increases.

Layers of underlying reality

Even before Shannon produced the abstract characters in the language of probability theory, which enabled sophisticated communications in general and computer communications in particular, there was the telegraph.(Shannon, 1948) The telegraph required an understanding of the underlying reality of electromagnetism to make transmission of information through a wire possible. The abstract characters used to interpret reality and to provide the channel of communication were Maxwell's equations of electromagnetism. (Maxwell, 1854) The equations required education in decoding of mathematics. Education requires a stable environment and the technology for a building for the children, for food, clothing and shelter. This is an example of decoding the underlying reality in layers.

The layers are: understanding how to make food, clothing, shelter, building, math. This yielded Maxwell. More layers are Maxwell's equations, engineers applying his equations, the technology of making communications equipment, and Shannon inventing more math. All these required understanding the underlying realities so the technology could be invented.

The layers of decoding the underlying reality to enable the telegraph are: the presence of copper in the dirt, smelting copper from the dirt, inventing the process to make wire from the copper, decoding the properties of electricity and magnetism, and decoding the mathematics of communications.

Recapitulating, starting at the layer when man recognized that an underlying reality of dirt included copper metal, man realized he could smelt dirt and get copper metal. Seeking the underlying reality of abundance, man investigated the underlying reality required to produce massive amounts of copper. The expression of the desire to multiply the products of earth such as copper yielded the mental concept of cooperating work gangs and the manufacturing process. To organize a manufacturing process, man used the discovery of the underlying reality of language as a communication system that represented heaven, man and earth as sounds. Then there were the layers of mental mathematical inventions, World3: the layer of math; the layer of electromagnetic math; the layer of probability math; and the layer of communication theory math necessary for coding and decoding secret messages and the math which enables computers.

This book has reduced the underlying reality to a communication system which includes:

a) The information intended for communication,
b) The process of encoding and transmitting the information,
c) The communication channel,
d) The encoded information transmitted through the communication channel,
e) The process of adding noise to the communication
f) The process of receiving the communication
g) The process of decoding the communication to yield information,
h) The effect of the information system on the world and on human affairs.

Figure 1 shows the diffusion of information as a Generalized Communication and Signal Processing System.

Herein, one can ponder the reduced sub-set of underlying reality using only the information and its communication system. The communication system through which the underlying reality transmits information is composed of the entire heaven, earth and man. This requires some assumptions and hypotheses about underlying reality and its relationship with heaven, man and earth.

Refer to Appendix A. Assumptions and Hypotheses Are Introduced to Reduce the Problem of Intercepting the Communications from the Underlying Reality and the Problem of Decoding the Information Received

One could assume that heaven is the underlying reality. Then heaven transmits the information about the way that earth and man take form, how they react, how they change over time, how they evolve, how they emerge from previous entities, how they exist, how they reproduce, how they die, how they decompose. The ancient Hindus named gods which sent the information for creation, preservation, and destruction to earth.

The influence of the underlying reality is not restricted to transmitting information to the physical basis of existence. There are many possible assumptions and hypotheses which would lead the investigation in different directions. One could assume that heaven is not part of the underlying reality. Assume the complete set of influences of underlying reality on heaven, earth and man is yet to be discovered. Assume that a sub-set of the expression of the underlying reality creates, preserves, and destroys heaven and that heaven is non-material; it exists as mind itself and fields emanated by mind. Assume that a sub-set of the expression of underlying reality includes the creation, preservation and destruction of the physical nature of heaven, earth and man, of the processes of physical things, of the diffusion of physical entities, and also the preponderance of the wave form in observations. This is a major problem; what to assume and what to hypothesize.

An ancient example of
wave motions, air flows and chaos

Chaos can be observed in most processes. (Bird, 2003)Near the banks of a stream one observes laminar water flow but in the center there is turbulent chaos. Even in the mathematical descriptions of chaos, there are patterns that almost repeat. These regions of similar but not exact repetition are called "attractors." The attractor appears to have a description that is a summation of waves of different frequencies and different amplitudes. The waves at an ocean beach have the form of a wave but each wave is different. Each wave is the summation of many waves contributed by the chaos of different frequencies and amplitudes.

Perhaps the shifting tectonic plates under the ocean have frequencies on the order of one wave per hundreds of years. (Bird, 2003) One could observe the cycle of rabbit over-population and starvation due to the placement of mountains, valleys, agricultural fields, the waves in a stream, and the frequency of rains alternating with droughts.

There is a specialty in mathematics called Fourier analysis that can describe some apparently chaotic phenomena. This analysis sums up the many component waves to yield the overall resulting wave. One can observe phenomena that are composed of processes that conflict with one another and later cooperate with one another. This is usually the source of perfectly repeating wave motion. This is also the source of nearly repeating wave motion in chaos.

For example, an expert architect or Feng Shui master could recognize the inherent chaos or regular wave motion in a region if landscape features were adjusted to aid the chaos or regular wave motion. Then adjustments could be made to prevent catastrophic chaos.

Would the concepts of waves and the instruments for measuring waves aid architecture and Feng Shui analysis and design?

The mental constructions called Feng Shui were the developed by a society founded on the teachings of Taoism, Buddha, and Confucius. These three systems proposed methods of understanding the underlying reality of heaven, man and earth.

The scientific system of understanding the world often employs wave forms as normative descriptions. Even before science became a dominating explanation, men examined wave motions. Feng Shui practitioners who had access to higher levels of mind sought to receive communications from an underlying reality. Would they have analyzed the landscapes, buildings, and homes comprehensively if they had wave concepts and instruments available?

Can the definition of underlying reality be expanded through principles of quantum physics?

There is a set of knowledge about the earth and how the forces of earth affected human life. Wind forces, water forces, stone forces, light, and so on are known. There is a limit to what influences can be included in the Feng Shui and architectural analysis and design. However, the Feng Shui practitioners and architects search for the content in the underlying reality that influences a building. The building is subjected to influences for the 100 years after construction is completed. The skill and reputation of these professionals depends on every possible tool, even quantum physics, to ensure the usefulness of the building.

Einstein wrote the math description of fluctuations of physical sub atomic events.

Einstein's description initiated pursuit of this mental construction ever since. The implication is that events could occur during extremely short times and extremely short lengths that would yield mass as electron-anti-electron pairs. There is a mathematical probability that another event during this short time could void the anti-electron and thus the mass of an electron would be created. Other entities, such as electromagnetic radiation, could also be created in the same way.

The science of physics has hypothesized a mechanism which would create particles from a field when there is no mass but there is energy, three dimensions of length, two dimensions of time, and a trigger to cause the fluctuating wave of energy to congeal into a lasting particle. This could be called 'creation from emptiness' except there is energy within the emptiness before the fluctuations yield matter.

Einstein began studying energy fluctuations in 1904. (Einstein, Albert, 1904, p. 354)

He derived the energy fluctuation formula in 1909. (Einstein, Albert, 1909, p. 185)

$<e^2>=KT^2 \ \partial<E>/\partial T$

Where $<e^2>$= mean square energy fluctuation

$<E>$=average energy for a system in contact with a hot bath at temperature, T.

K=Boltzmann's constant

$\partial<E>/\partial T$ means the change in energy as the temperature changes. (Pais, Abraham, 1982, p. 402)

Research into the concept of void or nothingness has resulted in the conclusion that there is no void. Physicists concluded that there is always some energy and other entities in a 4 dimensional worldline [three length dimensions and one time dimension.] For example, if one examines the electromagnetic spectrum of visible light emitted by a radiant heat source, then one finds that there is an ambiguity in the frequencies due to the fluctuations of the atoms of the heat source as they fluctuate between two states which could be potential energy and kinetic energy. See *Fourier Series and Optical Transform Techniques in Contemporary Optics* (Wilson, R. G., 1995) (Goodman, 1968)

Consider electron-positron pairs. Everything about an electron is exactly the opposite of everything about a positron. Each of the pairs are exactly opposites (Yin-Yang) so they when they collide, they exterminate each other. Thus the underlying reality, the void, can produce a fluctuation yielding an electron-positron pair. But it maintains the conservation of mass and energy since the mass and energy of the pair sum to zero. For a statistical collection of a large number of pairs, each pair exists for a different duration of time. The duration can be extremely short or can approach infinite time.

Many physicists have pondered whether the fluctuations could conceivably exhibit adequate energy and a trigger to initiate a matched pair of sub atomic particles. The theory of fluctuations is intended to address particles becoming persistent mass.

Many physicists agreed with the completion of the particle picture. A photon particle is a state of the electromagnetic field. It has the following properties.

1. Any particular photon has a definite frequency, nu, and a definite wave vector, $k \rightarrow$

2. It has energy E=h x nu where h is Planck's constant and nu is the frequency.
   The Planck constant of action has the dimension of specific relative angular momentum in joule-seconds or electron volts-second.
   The value of the Planck constant
   is h=6.62606957 x 10^{-34} joule-sec. = 4.135667516 x10^{-15} eV-sec.
   The Planck constant is related to the quantization of light and matter. It has the sub-atomic scale.

3. The value of its spin is one and it has two states of polarization.

4. It has no mass but it was experimentally discovered that the mass is less than 8 X 10^{-49} grams
   (Pais, 1982, p. 407) (Wigner, 1939, p. 149).

Heisenberg proposed the Uncertainty Principle from which could be inferred that when the order of magnitude of the observed sub-atomic process is less then 10-34 joule-sec then one can not know accurately about all participants in the process. This inference is stretching the meaning of the principle. However, the inference allows some room for scientific hypothesis. (Heisenberg, 1962)

The inherent uncertainty in decoding the underlying reality

There are many levels of reality. There are many levels of mind. There are many levels of consciousness. There are many more of these levels that almost no one experiences. However, humans habitually perceive the order in the chaotic environment. The perception of chaos is due to the interconnectedness of all processes and entities. Peat demonstrates the many human perceptions that are inherently uncertain in *From Certainty to Uncertainty: the Story of Science and Ideas in the Twentieth Century.* (Peat, 2002) The scientific professions and other systematic arts and trades construct a mental order out of the chaos.

No one can change the natural laws of light and mass. The fluctuating charges in the vacuum are also the reasons that light rays cannot simply penetrate each other without perturbing each other. And this is so for all wavelengths of electromagnetic radiation. (Genz, 1999, p.227-248 Chapter 7)

Recommended research into creation of mind

This leads to the question, "Does the summation of all the electromagnetic radiation produced by the human body yield the human mind which persists through duration of time?" This question is the research gate into the emergence of the mind out of the physical body. The gate will not be opened in this book.

Photons, a band of wavelengths of electromagnetic radiation, are conjectured to couple to the electric charge. Electric charges emit and absorb light. (Genz, 1999, fig. 61 page 206) All light effects are conjectured to be reduced to this elementary process. (Genz, 1999, Figure 78 page 248) This has not been experimentally proven or falsified as of 1999. However, the scattering of light in the strong electric field of an atomic nucleus was experimentally verified with gamma rays of very short wavelengths.

Let us recall that any action occurring in less then the Planck constant of action, about 10-34 joule-sec, exists in a condition that can not be determined or measured because of the Uncertainty Principle. Thus, this action could violate some of the physical laws governing actions in the realm of certainty and measurement. From this point of view, a principle is conjectured that a fleeting particle-antiparticle pair fluctuating in the vacuum can fuse into a permanent particle in the presence of a suitable field. This allows the fusion of a fleeting photon-anti-photon pair into a single permanent photon in the presence of an external field. This novel reality was constructed out of the quantum theory of wave fluctuations and the Uncertainty Principle applied to creating mass out of emptiness.

Recommended research into the psychic field

Can this conjectured principle be used as an analogy to answer the question, "Would a fleeting psychic-anti-psychic pair fuse into a single permanent psychic entity, such as Chi, in the presence of a persisting psychic field? The field can be emanated by one or more minds.

The origins of Chi out of nothing are considered

Discussion of Chi including proposed definitions, is provided in Appendix E What is Chi?

Hypothesis: Chi is created by the same mechanism that creates sub-atomic events and particles

These topics will be continued in Chapter 4: The Buddhist Search for the Underlying Reality was Based in the Mind-Brain World.

Chapter 3

The scientific search for the underlying reality is based in the material world

Abstract

The scientific search for the underlying reality is based in the material world. Can the methods for material research be applied to research applied to the brain-consciousness-mind world?

Is mathematics inherent in the underlying reality? To be most effective, all sciences, including physics, attempt to collect information from human senses aided by measuring instruments and from the underlying reality not accessible to the senses. This scientific method is succinctly expressed in mathematical functions and operators. Feng Shui and architecture must collect these two kinds of information to be most effective in producing designs that yield the maximum usefulness of a landscape, building or home. Math tools are integrated into these disciplines.

The initial scientific objective is to address the problem of collecting information from the underlying reality.

The knowledge of how to decode wave borne information and how to apply it to technology is extended by mathematics which is a language of analogy. Consider how research into quantum physics may increase the information available for experimenting with methods of intercepting the underlying reality. (Coulson, 1977)

What is the underlying reality of the universe from the point of view of some physicists? The physicists study matter in the void, ether, geometric theories, the dimensions of space and time, the metric field, the vacuum complexity, geometric theories, the dimensions of space and time.

Architecture and Feng Shui are partly science and partly art. Their effectiveness increases with higher precision of measurement and with the accuracy of the decoding of underlying reality. The duration of a building or landscape also depends on the accuracy of knowing the history of the construction site. Earthquakes? Typhoons? War?

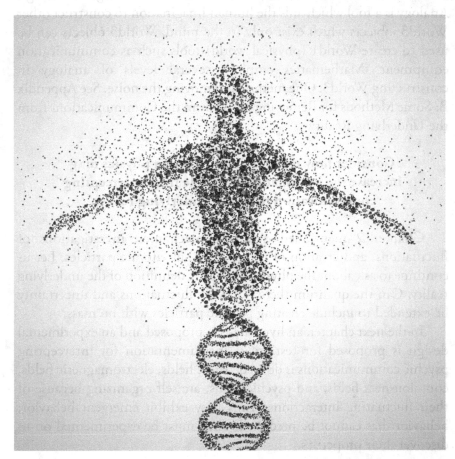

Buddhist practices also influenced the research into the underlying reality by Feng Shui practitioners.

Is mathematics inherent within the underlying reality?

The sciences have a foundation in math. The science of brain-consciousness-mind is less founded in math.

Gaining knowledge of how to decode wave borne information can be accelerated by mathematical tools. Math is a language of analogy. Analogy is a tool which aids the human imagination to construct other World3; objects which exist only in the mind. World3 objects can be used to create World1 (physical world) tools such as communication equipment. Mathematics employs several levels of analogy in constructing World3. One method is removing the noise. See Appendix B: Some Methods for Removing Noise from the Communications from the Underlying Reality

Consider how research into quantum physics may increase the information available from intercepting communication from the underlying reality.

Chapter 2 mentioned quantum mechanics, investigations of fluctuations, and creation of reality from sub-atomic particles. Let us continue to ask about the physical scientific definition of the underlying reality. Can the quantum theory of wave fluctuations and uncertainty be extended to include creating psychic particles with no mass?

In the next chapter, an hypothesis is proposed and an experimental design is proposed for testing it. Instrumentation for intercepting psychic communications is desired. These fields, electromagnetic fields, consciousness fields, and psychic fields, are self organizing because of their fluctuating interactions. Thus they exhibit emergent behavior; behavior that cannot be predicted. They must be experimented on to discover their properties.

Can the quantum theory of wave fluctuations and uncertainty be extended to include creating psychic particles with no mass?

This question was foreshadowed in Chapter 2. It is obvious that the mind interacts with its body. An unanswered question, "Does the mind interact with other parts of the physical world?" This question has inspired endless writing in the field of quantum physics. Research has suggested that subatomic particles can be generated by fluctuations in the underlying reality. A question in the field of consciousness is, "Can this research be applied to the generation of mind?" The question beyond the forefront of the study of mind and consciousness is, "Can psychic particles or psychic entities be created in analogy to the creation of subatomic particles?" One could answer this, "One does not know the entire process when the fluctuation has dimensions or entities on the order of 10-34 joule-sec"

Does the mind of the architect or the Feng Shui practitioner interact with the space he is analyzing? How can this be investigated? See the next chapter.

Scientifically rigorous instrumentation for intercepting psychic communications is desired

One problem with scientific investigation of human extrasensory perception is the lack of instrumentation. If there were ways to reliably transmit, record, receive, and preserve human psychic communications, then the psychic science would advance quickly.

In communication science, mathematics has always pointed the way to the technology. The key to finding the math, and the technology which describes extrasensory perception, is realizing that the psychic field is only partly an electromagnetic field. Some of the electromagnetic fields of the human body have been measured and described mathematically. (Gulrajani, 1998) These fields are used in medical monitoring equipment, *e.g.*: EKG, EEG.

Hypothesis: One probable model of the universe is an information processing system in which the inputs from the underlying reality are undetermined.

Every thing within our human ability to detect, such as field, atom, earth, process, nervous systems, is then converted into a binary item of information. (Vedral, 2010) Consider the noise added by the false but unfalsifiable human mental constructions. The process of science is then imagined as a decoding of the information inputs by underlying reality. How much is science corrupted by the human mental habit of inferring more than what is logical? How much has unfalsifiable human imagination participated in the description of the scientific observation?

Then one asks, "Is this observation the ground state, uncorrupted by human invention?" The answer is yes or no which yields to a binary system of math. (Wheeler, 1989)

One could study matter in the void, ether, geometric theories, the dimensions of space and time, the metric field, and the vacuum complexity.

To improve his analysis and design of an experiment to intercept the underlying reality, an investigator may exist who has attained the mind levels which will be suggested in Appendix D. An Attempt to Label Some of the Separate Levels of Mind and Consciousness. The mental states developed through Buddhist practice about concepts within the Buddhist mental environment will make an investigator capable of enhanced senses and Extra Sensory Perception (ESP). Such an investigator would be aware of more influences in the space under study than ordinary persons untaught in the enlightened mental environment.

Matter in the void: ether, space, and mind fields

Sir Isaac Newton laid the groundwork for the modern physical sciences in 1687. Before him, no one had tried to differentiate between the realm of initial conditions over which we have some control, and that of the laws of nature over which we have no control. He determined that after the initial conditions are set for a rock, the laws of mechanics will fix the rock's trajectory. (Genz, 1999, p.145 Chapter 4)

Space, Time and the Metric Field

We can characterize the curvature of space with angles in a triangle. When the sum of the three angles in a triangle equals 180 degrees then the curvature is zero and the space is flat. The theory of relativity refers to the metric field with a curvature in four dimensions, three lengths and one time. A four dimensional space-time contains fields such as a gravitational field. So it is not empty. Einstein's point of view was that the field defines the space and the masses in the space define the curvature. So space is a properly modified expression for the positional quality of the world of masses. If no masses, then no space. Gauss invented the "Gauss metric" to describe this frame of reference. (Genz, 1999, p. 173-178)

It is possible to define a metric field with three time dimensions.

The vacuum complexity

The concepts of a vacuum and a location with zero energy have been an object of study for many years. (Barrow 2009) (Dhammajoti and Dhammajothi, 2009) (Genz 1999)

Blackbody radiation at zero temperature has been studied as the Casimir effect which will be described below. (Genz, 1999, p. 189)

Those who have scrutinized quantum mechanics realize the limits to knowing the underlying reality

The geniuses who invented quantum mechanics realized that they could not know the true nature of the basic structure of matter. Many of them were open minded enough not to insist that their inventions were the underlying reality.

Much of physics is formulated in the language of math which is not intelligible by most people. For those not conversant in math, this obscures the concepts and the mental process of inventing the physics. But the objective of physics is to decode the underlying reality.

Hypothesis: the universe is an interconnected entity.

The following sections deal with ideas about breaking the symmetry of the vacuum. The ideas are named after the people who formulated them, Jeffrey Goldstone and Peter Higgs. First imagine the objects of interest, the ideas. One's imagination is based on one's experiences which limits one's imagination. Many of the mind objects in quantum physics are not in our experience, such as particles without a position. To clarify them, they are presented herein as analogs to our experience. (Genz, 1999, p. 257 Chapter 8)

Geometric theories and the dimensions of space and time

The general relativity theory has a geometric formulation of the gravitational force. It serves as an example for other theories of interactions. The concept is that the trajectories of masses are the shortest path between two points in four dimensions, not three. Kaluza and Klein expanded the theory of relativity to include the electromagnetic interactions. They introduced a second dimension of time which recognizes that there are more time dimensions than humans are normally aware of. There are many properties that humans cannot experience but we can try to decode the evidence which we can experience to infer some of the underlying reality such as quantum mechanics.

What is the underlying reality of the universe from the point of view of some physicists?

Edward Tryon conjectured a universe that is a vacuum that has fluctuations with positive and negative parts which sum to zero. Quantum theory predicts fluctuations with positive and negative virtual particles that lack enough energy to become real. Another conjecture is that there is some entity that has less energy than nothing. The conjectured vacuum loans large amounts of energy for short times and small amounts of energy for long times. On the one hand, if the energy approaches zero, the fluctuation that we call the universe may last for a time approaching infinity. But if the energy is zero or negative then the universe will last forever. Strong electric fields can produce electron-positron pairs of particles. (Genz, 1999, p. 261-262)

There is a belief or hypothesis that atomic scale particles have two types of mechanical energy, potential and kinetic. The particles vibrate between the two states. This was observed in the Fourier analysis of light transmitted through a crystal. The observed light waves and their interference patterns are a little fuzzy due to the vibration. There is also the energy, E, in a mass m, where the relation is $E=mc2$. In addition, the vacuum, with no energy or mass present, is hypothesized to have inherent energy.

The uncertainty principle in quantum mechanics leads to a hypothetical relationship of time and length. One of the mental constructions of most large groups of humans is the origin of all that exists; the creation stories. The science industry has its own creation story; fluctuations create matter. Consider the hypothesis: Fluctuations exist which create mass. These fluctuations occur in extremely short distances and extremely small durations of time, on the order of 10^{-34} Joule-seconds. $\sigma x \sigma P = h/2\pi$ equation (1)

Where σx is the standard deviation of the position, x; a measure of the uncertainty of the fluctuation. And σP is the standard deviation of the momentum, p; a measure of the uncertainty of the fluctuation. And h is Planck's constant which is divided by 2 and the universal constant π. The Planck constant, h is the limiting value in the Heisenberg Uncertainty Principle. Any action of which its momentum multiplied by position that occurs in less than about 10^{-34} joule-seconds is beyond the limits of certainty. Therefore on this scale, events, such as mass creation, could happen that violate physical laws at longer durations.

Consider that there are fluctuations that could yield electron and anti-electron pairs that would recombine in less than 10-34 joule-seconds thus not violate the law of conservation of energy. Richard Feynman used this hypothesis as a basis for his theory of interactions of sub- atomic particles. Then there is a set of events that occur at lengths below a certain threshold and that occur for durations of time below a threshold which would account for sub-atomic observations. R. Feynman received the Nobel Prize for this thinking.

Architecture and Feng Shui increase in precision in proportion to the accuracy of the decoding of underlying reality

Clearly the concepts of space, time, energy, and the underlying reality have changed considerably since Feng Shui was invented. The utility of Feng Shui is grounded on receiving accurate communications from human senses and from sources not available to the senses, extra sensory perception (ESP). Quantum physics treats communications not available to the senses and perhaps not available to known measuring instruments. These are the theoretical limits of knowing the underlying reality in physical science.

In contrast, the Buddhist search was based in pure mind. See Appendix D: An Attempt to Label Some of the Separate Levels of Mind and Consciousness

Chapter 4

The Buddhist Search for the Underlying Reality was Based in the Mind-Brain World

Buddhist methods of training the mind to be skilled in the perception of the underlying reality were taught since 500 BC. Feng Shui originated in concepts within the Buddhist mental environment grounded in purifying the mind. It is doubtful that Buddhist practice has ever been applied to architecture.

Consider the hypothetical architect or Feng Shui practitioner who has improved himself using Buddhist methods. Ponder the attributes he could employ to improve his professional applications to a high rise residential and business compound with complex initial conditions including mountains, rivers and hurricanes.

A man experiencing higher states of mind may receive communication directly without a sense experience.

One may use the many levels of the mind itself to locate channels of communication from the underlying reality. Consider the following training process recommended by Buddha to become purified from many influences on the mind. (Buddha, 500 BC)

Ten sources of sense data transmit information to brain and mind.

Begin with ten senses transmitting information to mind through the channel of consciousness. In most cases, the mind does not process all the information all the time.

0. Autonomic operations such as breathing and reflexes.
1. Mind receives smell signals.
2. Mind receives sight signals.
3. Mind receives taste signals.
4. Mind receives sound signals.
5. Mind receives signals from internal enteroceptors of touch, external exteroceptors of touch, and proprioceptors, knowledge of the locations of all body parts.
6. Mind receives memories from the mind and nervous system.

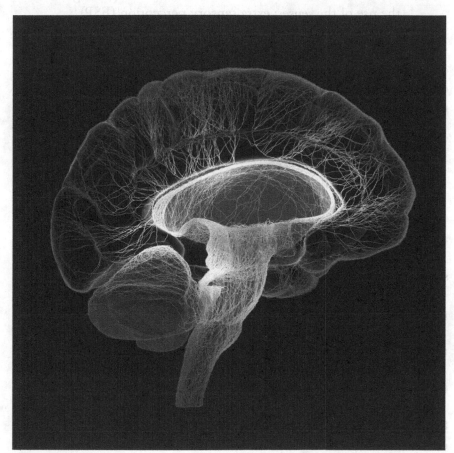

7. Mind receives awareness of one's unique aggregate of components constituting the ego.
8. Mind participates in cognition, intellectual manipulation, and creating mental objects.
9. Mind receives emotions and the coloring of thought, speech, and bodily actions due to emotions.
10. Mind receives communication from other living beings without the use of the senses; extra sensory perception (ESP).

ESP will be discussed in Chapter 8 Decoding the many channels of communication from the underlying reality requires a higher level of mind and consciousness of individuals and also of the group of researchers

Nine levels of mind

Let us explore the higher levels of mind.

Let mind level 0 be defined by non-conscious reflex and autonomic actions.

Let mind level, jhana 1 be seclusion and detachment from part of senses 0 through 5.

Let mind level, jhana 2 be stopping most reactions to senses 0 through 5, and stopping 8.

Let mind level, jhana 3 be stopping most reception of senses 0 through 6 and stopping 8, there is no craving, no clinging, no coveting.

Let mind level, jhana 4 be stopping all reception of senses 0 through 6 and stopping 8 and 9.

Let super mundane level 1 be the stopping of 0 through 6, stopping 8 and 9 senses and losing the limitations of space; yielding infinite space or no knowledge of finite space.

Let super-mundane level 2 be the stopping of senses 0 through 6, stopping 8 and 9, losing the limitations of space and losing the limitations of receiving information from only the ten senses; yielding infinite consciousness of everything. This enables ESP.

Let super-mundane level 3 be the stopping of senses 0 through 10, non-attention to expansion of space and consciousness; yielding void, emptiness. This is the state when the ego, the grasping at one's self, disappears. ESP is no longer sensed.

Let super-mundane level 4 be the stopping of senses 0 through 10, non-attention to expansion of space and consciousness. There is pure Being based in emptiness, neither non-perception nor perception. In this level, one does not accumulate karma, one is liberated from the cycle of life and death. There is no craving, no clinging, and no coveting. See Appendix D: An attempt to label some of the separate levels of Mind and Consciousness.

It is obvious that a man in the super-mundane level 4 can return to lower levels of mind from which he will perceive cause-effect processes and probable events. He will be aware of the limitations of predictable events, limitations on his 11 senses, and limitations on his ability to comprehend all that is happening everywhere. These perceptions will include the factors that architecture and Feng Shui seek to identify: measurements of space and time, cyclic influences in the past and future such as weather, influences and changes of Chi in animals and plants, the effects of mirrors and colors, the ineluctable influences on the eleven senses of a human and so on. The architecture or Feng Shui practitioner would have a vast increase of information about the space being investigated. He would have an abundance of understanding of present and future events based in causality. He could conjecture the effects of probable influences on present and future events. He would not try to cope with most unpredictable events unless they are catastrophic.

See Appendix D: An Attempt to Label Some of the Separate Levels of Mind and Consciousness

After the realization of all eleven sensory conditions and the nine higher levels of mind, an investigator would experience a different ordinary reality which would be enhanced by intuition about several layers of underlying reality. There will be fruitful research for the most fundamental level of underlying reality, which may be emptiness, and for imagining how to detect and record the communications from the fundamental level.

Without training in the above nine levels of higher mind and the subsequent skills, what are the limitations on the practitioner? Considering the inherent potential to bring into reality various levels of mind and consciousness, would the arts and sciences of architecture and Feng Shu become polished to perfection by incorporating this training into the standard education of these professions?

Consider the mental states such practitioners could achieve described above.

Perhaps it is time to update architecture and Feng Shui principles to incorporate the ancient concepts of higher levels of mind and extra sensory perception to enhance the analysis and design developed by these professions.

Recommended research into the brain, consciousness, mind, and psychic field

I recommend that contemporary investigators of underlying reality enlist in this Buddhist training to increase their skill and knowledge. Countless populations have practiced Buddhist principles for living including concentration, meditation and absorption. Now the time is appropriate for researching the brain, consciousness and mind. The research will accelerate human group behavior beyond contemporary self centered motives focused on material gain.

Chapter 2 introduced the hypothesis from physical science of creation of mass through the mechanism of fluctuations and a triggering energy.

Can this conjectured principle be used as an analogy to answer the question, "Would a fleeting psychic-anti-psychic pair fuse into a single permanent psychic entity, such as Chi, in the presence of a persisting psychic field? The field can be emanated by one or more minds.

The creation of Chi out of nothing is considered

Discussion of Chi, including proposed definitions, is provided in Appendix E What is Chi?

Hypothesis: Chi is created by the same mechanism that creates sub-atomic events and particles

There is a probability of a given event occurring provided that the conditions are such that probability theory applies. Let M be the set of such fluctuation events which could yield Chi. Some of the events would yield mass, some would yield radiation, and some would yield

Chi. Others in the set would yield nothing. There are probabilities for yielding and for non- yielding. There is a probability wave that describes the yield of Chi. If the probability is less than zero, no Chi is created. If more than zero, then Chi is created. The sources of the probability wave are all the events that establish Chi and that allow the Chi to vanish before the short duration of the Chi event terminates.

The creation events are assumed to occur at random times in a chaotic environment. The duration of such events is less than the limit of the Uncertainty Principle but the durations are random. The sum of the probabilities of all the random events in the set M is the integer, 'one.' The sum of all the fluctuations during the set M could be described by a Fourier analysis.

The set M is transmitted into this world by the underlying reality. The transmission must have chaotic elements or else the fluctuations would not exist. In other words, if the transmission were a perfect frequency or set of frequencies the there would be no fluctuations that create mass, or Chi. There must be some slowly varying change in the transmission so that fluctuations would occur in the region below the limit in the Uncertainty Principle. No such fluctuation can occur above the limit. Fluctuations above the limit are excluded as defined by the Uncertainty Principle.

I recommend research into vitality, Chi

This mental construction of the creation of Chi in parallel with the creations of mass and electromagnetic radiation, is a novel idea. I recommend experiments to detect the properties of fluctuations yielding Chi. Experiments have been conducted to observe the fluctuations yielding mass and radiation. These could be used as a starting approach to research Chi. Fluctuations may be the source of Chi.

Clearly there are many difficult problems in defining the underlying reality, discovering where and when to intercept its communications, intercepting its communications, and decoding to yield useful information. The difficulties make the challenge to produce meaningful research on Chi more exciting.

Experimenting on the brain-consciousness-mind and measuring it are natural undertakings for a Buddhist practitioner.

An hypothesis to verify or to falsify and an experimental design for testing the hypothesis

Hypothesis: Psychic information is transmitted through a channel of particles without mass.

Experimental design

These hypotheses are proffered to stimulate experimental designs to verify or falsify this hypothesis. The most difficult problem in experimenting on the brain-consciousness-mind is to identify a channel of communication to monitor. The best channel is one that is used by the underlying reality.

An experiment is recommended to yield the following information.

a. Define the properties of psychic particles.
   Particles without mass can move at the speed of light or faster without violating the premise of the theory of relativity.
b. Define the effects caused by psychic particles.
c. Define the probable influences of psychic particles on the mind, human senses, and non-living physical material
d. Decode the properties of the psychic particle based on what is discovered about them.
e. Propose a non-physical or physical instrument to detect psychic particles.
   For example, certain humans can detect psychic particles.
f. Propose a means of generating as many psychic particles as desired.
g. Propose experimental conditions to detect the causal and the probable influences of a psychic particle, or a stream of particles.

Detect and measure the energy of the particle
Detect and measure the psychic field
Detect and measure the causal effect of the psychic particle and the probable influence on physical objects. (Genz, 1999) (Wigner, 1967)

The world is complicated and impossible for the human mind to understand completely. But there are accidents which allow a simple experiment to yield completely understood information about the world. One must seek an accident in the form of an exactly controlled experiment. This is a reduction of the complexity of the world in which one abstracts a non-accidental domain from the total interconnection of everything. In the abstraction, simple laws may be found. The complications are called initial conditions. The domain of regularities, not accidents, is called natural law. Such a division of reality has limits but it allows the natural sciences to produce results. This same approach will yield the parameters of creating the psychic field and the psychic particle with no mass.

Hypotheses are helpful in designing an experiment and in designing detection equipment for the experiment

There are many simplifying hypotheses and assumptions which increase the fruit of experiments

Hypothesis: The mind is composed of a summation of electromagnetic waves, consciousness waves, and psychic waves.

Hypothesis: There are electromagnetic fields, consciousness fields, and psychic fields.

Hypothesis: These fields, electromagnetic fields, consciousness fields, and psychic fields, are self organizing because of their fluctuating interactions.

The fields emerge from the fluctuations in the physical world as discussed in electromagnetic theory. (Maxwell, 1965)

The communication channel within the complex field composed of the psychic field, electromagnetic field, and consciousness field may be the superior method of identifying the communications channel from the underlying reality. Before the math can be fully developed, the psychic scientists must frame the taxonomy, the vocabulary, and the psychic tools for the task of identifying the message from the underlying reality. Using these mental constructs, one could distill the message from the underlying reality that is not corrupted by noise.

The characteristics of the human senses and the human faculties for extrasensory perception must be considered. The theories suggested in this paper can be applied to these senses and faculties.

Chapter 5

Communication from the underlying reality is defined

Abstract

Communication is generally composed of the components in Figure 1. A signal is the variation of a quantity by which information is conveyed regarding the state, the characteristics, the composition, the trajectory, the evolution, and the course of action or the intention of the information source. The information of interest is the past, current and future conditions and the influence of the underlying reality on the world. A signal is a means of conveying information regarding the states of a variable. How are the communications decoded? How is the meaning of the information in the signal defined?

Diffusion, a universal principle in all phenomena and matter, could be channel of communication.

There are many levels of communication and meanings.

Hypothesis: the variables within the information from the underlying reality include the natural laws, the wave form of communication from the underlying reality to the world, the human mental archetypes, the human instincts, the local creation of mass, energy, length, time, the harmony of the underlying reality with the world, and the chaotic conditions necessary for the emergence of life.

Hypothesis: the variables within the information from the underlying reality are the boundary conditions at which one can begin experimental investigations.

None of these variables, these boundary conditions, has been subjected to experiment with the specific intention of identifying it as a principle of the underlying reality.

The analogy of electromagnetic signal processing

Refer to Figure 1. The information conveyed in an electromagnetic communication signal, Info(t), may be used by humans or machines for communication, forecasting, decision-making, control, geophysical exploration, medical diagnosis, forensics, investigation of the underlying reality itself, etc. The types of signals that signal processing deals with include textual data, audio, ultrasonic, subsonic, image, electromagnetic, medical, biological, financial, seismic signals, establishment of material world, continuous maintenance of the material world, inventing the natural laws, undetermined nature of events, etc. Figure1 in Chapter 2 illustrates a communication system composed of

a) an information source, I(t),
b) followed by an encoding system, T, for transformation of the information into variations of a signal, x(t),
c) a communication channel, h, for propagation of the signal from the transmitter to the receiver, additive channel noise, n(t), and
d) a signal processing unit at the receiver for decoding (extraction of the noiseless signal and the information from the received signal).

In general, there is a mapping operation that encodes the output, I(t), of an information source (including the underlying reality as a source) of the signal, x(t), that carries the information. This encoding operator may be denoted as T·and expressed as

$$x(t)=TI(t) \hspace{4cm} \text{equation (2)}$$

The information source, I(t), is normally discrete-valued, but the underlying reality could take on unknown characteristics. The signal, x(t), that carries the information to a receiver may be continuous, discrete or other as yet unknown characteristics. For example, in multimedia communication, the information from a computer, or any other digital communication device, is in the form of a sequence of binary numbers (ones and zeros) which would need to be encoded and transformed into a wave form, voltage or current variations, and modulated to the appropriate form for transmission in a communication channel over a physical link.

As a further example in human speech communication, the voice-generating mechanism provides a means for the speaker to encode each discrete word into a distinct pattern of modulation of the acoustic vibrations of air that can propagate to the listener. This includes tone, intensity, accent, emphasis, emotion, respect or offense, background noise, etc. To communicate a word, w, the speaker generates an acoustic signal realization of the word, x(t); this acoustic signal may be contaminated by ambient noise and distorted by a communication channel, or impaired by the speaking abnormalities of the talker, and received as the noisy, distorted and incomplete signal y(t). This is modeled as

$$y(t)h[x(t)]+n(t) \hspace{4cm} \text{equation (3)}$$

In addition to conveying the spoken word, the acoustic speech signal has the capacity to convey information on the prosody (i.e. pitch, intonation and stress patterns in pronunciation) of speech and the speaking characteristics, accent and emotional state of the talker. The listener extracts this information by processing the signal y(t).

Compared to this processing of the wave form, consider how complex or how extremely simple the communication from the underlying reality could be. It must create, maintain, and terminate all the phenomena of the world. To be completely scientific, one cannot assume what part the underlying reality takes.

Finally, after the noise is subtracted, the information is decoded. Before it can be used for human action, it must be given meaning.

The final decoding of the information is to give it meaning. Often the information appears to be nonsense or to be lies. There is meaning based on context, on function, on how well the information is known, and on the subset of reality of the person who uses it. In law courts, the meaning of the words and the arguments are determined by the theory of interpretation, on intention, on the morals of who presented the arguments, on precedents, and on the values of the judge. (Moore, 1986)

Diffusion is a possible channel of communication

Diffusion is a pervasive phenomenon found in all processes and materials. There are endless books about the science and engineering of diffusion processes. Most major industries use diffusion in the processing of materials into the final product. Ideas diffuse through consciousness into the mind of man.

For example, the Feng Shui inventors were, highly educated and free to ponder diffusion in daily life. The total building environment diffuses into the mind of a Feng Shui practitioner.

In Feng Shui the key to decoding the properties of underlying reality is discovering whether the underlying principles such as wave forms, diffusion, and the laws of physics or the natural laws of justice are themselves the underlying reality. Are they the constituent structure of the underlying reality? Or does the underlying reality communicate these World3 instructions about how to give order to objects in the material world? How does one define the difference between the underlying reality and its expressions in the world?

One could conjecture that the Feng Shui practitioners realized that air, heat, water, Chi, dust, and ideas diffuse through a residence or a working factory. The things that diffuse and can be understood as waves are included in factors that require a harmonious treatment. (Howard, 2012, Chapters 13, 14, 15)

Chapter 6

Many different varieties of reality were discovered in attempts to satisfy the curiosity about the underlying source of reality

When do humans begin to search desperately for reality?

In all of human history, there have been an endless number of attempts to know the attributes of the underlying reality. These efforts usually resulted in hypotheses that could be called religious dogma, common sense or traditional explanations of reality. One may not have decoded all the communication from the underlying reality but that does not mean it will never be decoded. Thus one must prepare in advance how to approach the higher level decoding process. This book recommends specific approaches and experimental methods to intercept and decode communications from the underlying reality.

Attempts to satisfy the curiosity about the underlying source of reality

Popper defined the term, "Real." It means material things of ordinary size. This is extended to things that are too big or too small to grasp fully. They exert their existence as a cause of the effects on other things." The hypothesis of the cause and effect must also be accepted to believe a thing is real. (Popper. and Eccles, 1977, Chapter P1, p. 9-10)

Popper defined three worlds of reality, World1, World2, World3. World1 is the world of physical entities, things or states. World2 is the world of mental states which interact with our bodies, our consciousness, and our disposition to act. One's mental state causes one to avoid walking into a wall. World3 is contents of thought and the products of the human mind such as knowledge, arts, crafts, professions, and inventions. World3 can cause a man to build a house, machine or to cure a disease. (Popper and Eccles, 1977, Chapter P2, p. 36-50)

Popper criticized materialism in its many philosophical renderings. Popper explained the following realities: Physicalism, Radical

Materialism, Panpsychism, Epiphenomenalism, Identity Theory, and Parallelism. (Popper and Eccles, 1977, Chapter P3, p. 51-99) This is a partial list major groups who are convinced of the existence of different underling realities.

See Appendix C: Many Different Varieties of Reality. It discusses a few of the great number of decoding methods of reality accepted by large groups of people.

Many people scrutinize their world looking for patterns, themes, hypotheses, or assumptions about the underlying reality which they can use for their advantage. They examine human behavior, weather, car traffic, electronic laws and so on. They look in many channels of communication. There is a persisting effort by many humans, not just by scientists, to decode information from the underlying reality.

The scientific answer to these approaches

One cannot assume what form the communication takes or what is the total communication or what in the world receives the communication. The scientific method requires formulating hypotheses and testing them. A higher level of mind is required to employ the scientific method in the task of intercepting and decoding the underlying reality.

Lao Tzu decoded the Tao

Higher mental levels enable a man to grasp the methods of intercepting the communications of the underlying reality. An example of a man who evolved a higher state of mind was Lao Tzu. He was able to dispense with distractions and mental disturbances to yield a clear analysis of a decoded Tao. His analysis is embodied in the *Tao Te Ching*, translated to mean The Way of Virtue Text. According to tradition, it was written around the 6th century BC by the sage Lao Tzu, "Old Master" a record-keeper at the Zhou Dynasty court, by whose name the text is known in China. The text's true authorship and date of composition or compilation are still debated. (Kohn and LaFargue, 1998).

The passages are ambiguous, and topics range from political advice for rulers to practical wisdom for ordinary people. Because the variety of interpretation is virtually limitless, not only for different people but for the same person over time, readers need to avoid making claims of objectivity or superiority about a given verse. Consider how to interpret the following excerpts. A message decoded from the Tao is Ineffability.

> The Way that can be told of is not an unvarying way;
> The names that can be named are not unvarying names.
> It was from the Nameless that Heaven and Earth sprang;
> The named is but the mother that rears the ten thousand creatures, each after its kind.

Lao Tzu may have described a state of existence underlying the universe and before the identification of time or space. This is an alternative to the scientific Big Bang. Another attribute of Tao is Mysterious Female.

> The Valley Spirit never dies
> It is named the Mysterious Female.
> And the doorway of the Mysterious Female
> Is the base from which Heaven and Earth sprang.
> It is there within us all the while;
> Draw upon it as you will, it never runs dry.

Another feature of Tao is easy to observe in all things: birth, process and death. The verse is titled, Returning of Union with the Primordial.

> In Tao, the only motion is returning;
> The only useful quality is weakness.
> For, though all creatures under heaven are the products of Being,
> Being itself is the product of Not-being.

Lao Tzu decoded the Tao to have the property of emptiness.

> We put thirty spokes together and call it a wheel;
> But it is on the space where there is nothing that the usefulness of the wheel depends.
> We turn clay to make a vessel;

But it is on the space where there is nothing that the usefulness of the vessel depends.
We pierce doors and windows to make a house;
And it is on these spaces where there is nothing that the usefulness of the house depends.
Therefore just as we take advantage of what is, we should recognize the usefulness of what is not.

Emptiness has been an important subject of math and experiment for the last 100 years in the physics profession. Emptiness was also decoded by the Buddha who taught a method of realizing the void as one's being. See Appendix C.

Philosophical vacuity is a common theme among Asian wisdom traditions including Taoism. Especially *Wu Wei,* spontaneous effortless action sort of like a vacuum. One could decode the *Tao Te Ching* as a suite of variations on the "Expressions of Nothingness." In Buddhism, there is a phrase, "form is emptiness, emptiness is form" Emptiness, void, nothingness and vacuum were mentioned above and will be explored below. (Dalai Lama, 2012, p. 147ff)

Chapter 7

Many layers of instruments and mathematics were invented to overcome the limitations of human senses

The researcher must receive all the information, perhaps the future probabilities of change for the phenomenon. What are the limitations of receiving all this? Consider only the human senses that are limited to receiving certain types of wave representations of the environment: sight and hearing. Stimulation from the human sense organs is part of an information communication system. Information is transmitted by some entity in the environment. The whole information communication system is composed of the original underlying phenomenon, the encoding of the phenomenon into a signal, transmission to the human senses, addition of noise, transmission through a channel in the body, reception of the signal within the nervous system, decoding the signal, and delivering it to the conscious human mind and body for use. This is shown in Figure 1. This system is currently being studied intensely as information theory by perhaps a million engineers. (Luenberger, 2006) (Ash, 1965)

Human senses are more limited because only a tiny amount of all sense data are within consciousness, limited due to mental disorders, limited by emotionally caused misinterpretation, limited by beliefs, and limited because a human can only perceive what he is prepared to receive. See Appendix C: Many Different Varieties of Reality, in particular the *Diagnostic and Statistical Manual of Mental Disorders*.

Consider how humans learn to perceive through the senses. Each person creates a model of the environment based on sense data and on certain abilities inherent in the human. The abilities are inadequate to receive the total phenomenon, inadequate to decode the signal, inadequate mental faculty to create a valid model, extremely limited mental abilities to use the decoded information in the best way, and inadequate physical ability to respond to the signal. Since the human is so limited, the human has an abbreviated understanding, even a false conviction about the phenomenon that is transmitting the waves into his senses.

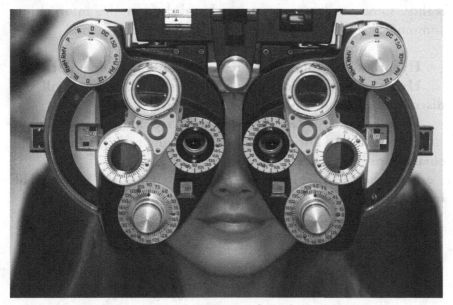

Therefore that original phenomenon that is transmitting waves is known from the fragments of the original sense data. The original transmitter can be labeled the ultimate source of reality. It is also labeled God, Brahman, Tao, and so on. Some of these labels are given attributes. The concept, 'underlying reality' has the least assigned attributes.

Excluding most of the attributes assigned arbitrarily to the unknown phenomena in the environment by the humans, let us consider what could be known about the source with the least attributes, underlying reality.

A great number of assumptions and hypotheses can be proposed and tested to reduce the difficulty of the problem of investigating reality.

However, even with limited abilities to discover the underlying reality, some attributes could be decoded from the information communicated by the environment. Some hypotheses follow.

Hypothesis: the underlying reality is nothing or does not exist.

Hypothesis: the underlying reality is a collection of all the discovered and undiscovered laws of nature.

Hypothesis: the underlying reality can not be known by definition.

Hypothesis: an attribute of the underlying reality is that all the emanations from it are interconnected.

Hypothesis: the underlying reality itself has the property of interconnectedness.

There are many similar initial conditions and points of view.

Historical methods of describing the information communication from the underlying reality to the environment based on minimum assumptions

There are an enormous number of beliefs held about the underlying reality. See Appendix A. Assumptions and Hypotheses Are Introduced to Reduce the Problem of Intercepting the Communications from the Underlying Reality and the Problem of Decoding the Information Received. Many methods have been employed to interpret the information in the communications. The recommended objectives are to test the hypotheses and to make calculations using the assumptions to find out if they give valid results.

The science industry has searched the underlying meaning and laws of the physical non-living and also the living world. There exists an enormous number of separate researches into nature. Consider these searches to be means of decoding the underlying reality.

An example of a recurring property of phenomena is the wave. Any fluctuating phenomena can be described mathematically with wave equations. One descriptive method is Fourier analysis, a mental World3 tool which has become an endless source of research as well as being embodied in electronic analysis tools.

Hypothesis: wave components of the underlying reality are commonly observed but not recognized.

Since waves are so common, one could assume that waves are part of the decoded communications from the environment.

How can one start with wave phenomena and use it to determine the attributes of the underlying reality? Are waves inherent in the underlying or are waves an encoded command to material things?

Another common element of the world is interconnectedness. The predominant property of waves in matter and in processes may be part of the mechanism of interconnection of many things. The above questions and simple examples demonstrate a few of the ways people decode their individual perceptions to arrive at a conviction of reality that they can live or die with.

Analogies between two or more phenomena are common.

Analogy can be used effectively as a teaching tool. Analogies aid the imagination to conceive of novel points of view. Some analogies demonstrate the wave motion and the interconnected nature of reality. Analogies stimulate the mind to satisfy the curiosity. Many analogies are presented below in various chapters. The analogy of the cello. The analogy of the bronze bowl: an experiment in decoding the expression of underlying reality. The analogy of the vibrant textile. The analogy of the financial news. The analogy of spying on electronic communications. The analogy of higher states of mind receiving communication directly without a sense experience. The analogy of the rock on the mountain.

The analogy of the spider web is an aid to decoding the communication and properties of the underling reality.

The analogy of the spider web is an aid to decoding the communication and properties of the underling reality

Some people realized that they could turn their perception at a right angle to introspect their feelings and intentions. This is called self examination. With training, one can turn one's perception at a right angle to one's usual experience of the moment. Then one can observe a longer continuous stream of sense data as a whole process within duration of time in one frame of mental perception or one vision. This is the second dimension of time, t5.

This is similar to a spider in the center of a web using the web as a sense organ sensing the world. The web becomes the world. The spider is a simile of a human trying to decode the signals within reality. If there is bug moving in the web, the bug is the underlying reality; the source of the signals transmitted by a bug in the web. This is the simile of a human sensing heaven, man and earth. One could again turn perception at a right angle to invent another analogy.

The second variation of this analogy would place the principles embodied in the underlying reality as the source of signals in the web. The underlying causes the bug to get tangled in the web.

The third variation of this analogy would be that the underlying reality is the buzzing of the bug which signals to the spider. The underlying reality could be assumed to be detected by humans in the form of signals. Signals represent the underlying. The properties of the material world limit the expression of the underlying. Man is aware of man, earth and all under heaven. He does not usually interpret this as expression of the underlying reality. Thus, in most cases, Man is not aware of the transformation of the underlying reality into signals that result in heaven, man and earth.

Hypothesis: the properties of the inherent makeup of the underlying reality would manifest themselves within the web, signals, spider, and also the earth, man and heaven.

Hypothesis: the underlying reality manifests itself within the observing human, the human senses, human nervous system, human mind, the human thinking processes of the mind, and the analogies invented by the human mind.

The fourth variant of the spider web analogy is that the spider is the underlying reality, creating the web of the environment, sending signals, vibrations, chemical messengers of paralysis and death to the bug, sending mechanical movements as wave motions, as vibrations, as fluctuations within the underlying chaos of the web. These are manifestations of underlying reality, measurable signals, measurable effects, chaotic attractors, entities which are stimulating to the spider's senses.

A method of decoding the underlying reality is to deny its existence

Appendix C discusses several methods of interpreting reality. It presents simple examples that demonstrate a few of the ways people decode their individual perceptions to arrive at a conviction of reality that they can live or die with. Examples will raise significant questions but will not answer them.

The investigator of the underlying reality must be able to understand the many analogies and hypotheses of underlying reality. How many investigators advance to this understanding?

In any case, the human senses are channels of communication leading to chakras, to synaptic nodes, into the solar plexus and other chakras, into the brain, stimulating the generation of the psychic field, the foundation of mind. This total experience is part of a multiple layered process of decoding the underlying reality. If the investigator knows this, can he discover the communications from the underlying reality?

Analogies follow which will raise significant questions but will not answer them. The following analogies demonstrate the wave nature and the interconnected nature.

The analogy of the cello

The fluctuations of the underlying reality may yield temporary entities. An expression of the Tao can be represented as waves. The significant sense data are types of air wave motion such as sound and electromagnetic sensations such as light. For example, a stringed instrument has strings of a certain length. Sound is produced by the wave motions of air produced by the strings. The musician constructs waves in the strings that are an exact multiple of the length of the strings.

Other waves sound terrible. This is a principle of many phenomena in our world. They are called standing waves. These have been found in many events, even in a vacuum. This will be noted in the bronze bowl and the Casimir Effect below.

All the parts of the cello performance are interconnected including the human musician, the physical environment constructed by humans, the electric lighting power and the heating system.

This standing wave behavior in cello strings is also hypothesized to occur in electron-positron pairs. The pairs are exactly opposites (Yin-Yang) so they when they collide, they exterminate each other. Thus the underlying reality, the void, can produce a fluctuation yielding an electron-positron pair but yet maintain the conservation of mass and energy since the mass and energy of the pair sum to zero. For a statistical collection of a large number of pairs, each pair exists for a different duration of time. The duration can be extremely short or can approach infinite time. A model for existing entities in accordance with physical mechanics is that there exists a single atom composed of several subatomic particles. These particles and their energies can only take on certain quantum values in analogy to the certain lengths that produce standing waves in a cello string. The values of the energy quanta in subatomic particles are the analogy of standing waves.

In the experimental apparatus for the Casimir Effect, an electron cannot exist without a matching positron. There is tiny space between two plates which are separated the distance of a standing wave of the virtual subatomic pair. Everything is interconnected. A small percentage of virtual particle pairs produced by fluctuations become real articles with momentum. Every creation of momentum has an equal and opposite momentum which can be measured. The momentum was created out of emptiness.

The analogy of the bronze bowl: an experiment in decoding the expression of underlying reality

Consider a bronze bowl of dimensions such that standing waves can be generated. This is often performed at a party using a campaign glass. Fill the bowl, or glass, with water and rub the edges of the open top with fingers or other substance that induces surface wave motions. This is more conveniently done with a thin Champaign glass; pushing a finger around the rim. The frequency of the vibrations induced by friction results in standing waves in the water surface. One can infer that the waves are a balance of forces, gravity, cohesion and so on. The standing waves are an average of the sum of forces and the sum of many waves. There were no waves initially and no mass was added or subtracted but waves were created. The wave laws, other causes, and other laws resulted in waves. The law of resonance within the wave laws induces the waves to be standing waves. Other causes will stop the waves when the energy from friction is withdrawn. This was created out of nothing and retreats to nothing.

One could decode the Tao in this experiment. The Tao communicates obedience to the laws of conservation of energy and usually the law of conservation of mass. After a process dissipates, the Tao communicates a return to the average state which existed before the process began. If the water is analogous to the encoded Tao, processes can manifest and the conservation laws are obeyed. The Tao emanates a reality where a small perturbation can initiate a process. A process can emanate out of the Tao due to a small perturbation and dissipate back into the Tao. If the Tao were within everything, then the potential for waves to exist are inherent in the Tao, not created from outside the Tao. This analogy demonstrates how standing waves appear in the uniform and featureless Tao out of nothing.

The analogy of the vibrant textile

Consider a woven textile. It can be viewed in six dimensions, length or warp, L, width or weft, W, thickness, T, the immediate time, tNOW or t4, the long view of all immediate changes, tLONG or t5, which could be called historical change in the long view of major influences, and probable changes due to the total of all influences, tPROB, t6. The textile initially has perfect horizontal form with right angles at the intersection of the warp and weft of the threads.

First, consider only L, W, T, and t4, as one holds the textile across the width in two hands. Let the textile be a kilometer long and let there be a frame of reference from which the locations of any part of the textile can be measured. One could move parts of the textile to see the changes in location of the remaining parts of the textile as time, t4, increases. One could stretch it to see how the warp and weft intersections change from right angles. In any case, all the threads are connected and influence each other.

Next, consider observing, t5 in addition. One could hold the end of the textile a kilometer away. Then one lifts and drops the textile in a rhythmic motion to see a wave form move toward the loom along the textile. The wave moves along the length, L of the textile. One would then anticipate the wave, or history of the textile. If the textile were to be analogous to any complex event such as the ocean, then one would realize that all the parts of the ocean are connected. This five dimensional view could be extrapolated to include any physical entity.

Next, consider that there are influences in a particular thread of the textile that could not be predicted such as the wear transmuting into a broken thread. The influence of the thread would migrate along the textile and result would be unpredictable. However if there were 1000 textiles that were manipulated, a random thread would break. This breaking event would be a random variable. A probability table could be written to show the occurrence of breakage. This is a record of probable time, tPROB, t6. This textile motion is a six dimensional phenomenon, three lengths and three times.

All elements of the textile and the environment are interconnected. This is the assumption that leads to decoding the interactions of the textile and to understanding the unknown influences on the textile.

One could investigate this manipulation, the resulting wave, the unpredictable behavior, the causal factors and the probable factors to yield an expanded grasp to any physical event such as the invention of a better loom.

Consider that the left hand corner of the textile is attached to a rigid support. Only the region, R, near the right hand corner is available to the human senses. How would one discover the length, L, width, W, thickness, T, and the changes that could be expected and predicted to occur in time? There are three times, tNOW, t4, immediate change, time tLONG, t5, and the time in the table of the random variable, of unpredictable events with the random variable time, tPROB, t6.

This is the simplified analogy of the problem which the scientific method attempts to solve. One could take the extreme generalization of this problem which is to discover L, W, T, and the changes in the textile as it is woven. The extreme is analogous to discovering characteristics of the underlying reality or the effects of the underlying reality on the world. In this case, the underlying reality is in all the hidden processes which are beyond the region, R which humans can sense.

The ancient Chinese, Confucius (551-479BC) may have explored this analogy of the textile. He wanted to explain the difference between acceptable behavior, when all lines of life (threads of textile) walked by men end in convergence with the total fabric of society, and deviant behavior, when one man starts his walk away the usual line of life (disciplined thread of textile). The deviant man diverges from the line textile, and ends at a different point from most others. He will be completely off the textile.

Another warning from Confucius is that when one man starts to work on a single strand of the fabric, he destroys the whole textile. Confucius concluded that deviant perspectives do not originate from divergent comprehension of the whole textile. They originate from trying to make absolute tendencies that only consider a single limited view of the whole. The single limited view originated from limited life experience (such as those who lived only in the limited reality of the Emperor's friends) or from experiences differing from most people (such as a merchant). These deviant views were based on a failure to have insight into their limited experiences compared with the whole balanced view of Chinese civilization.

The undecided concept of whether deviancy originates from moral limitations or out of intellectual single limited views remains today as part of the unsolved problem of Chinese thought.

An example is one of the four "represents." It is, 'The Chinese Communist Party (CCP) represents the people.' But simultaneously each top representative member of the CCP has a single limited view. Just as Confucius observed, he selfishly uses his connections to gain material possessions for himself at the expense of the whole textile of society. He ruins the collective textile of society by clinging and conniving to get influence for himself while destroying the influence of others. He gives prestigious positions for his family who are mostly incompetent to perform necessary functions for the collective textile of the city or the prefecture.

The analogy of the financial news

Michael Lewis described the detection of signals from the interconnectedness of finance with many activities of a global human population in *Panic: The Story of Modern Financial Insanity*. (Lewis, 2009) This is a more complex simile which is an approximation of all reality emanated from the underlying reality.

Consider the exchange rate between the US dollar and the Euro. It is not just a number; it also affects one's life by changing the cost of goods. It also indicates the productivity of American workers compared to European workers. The ratio of the currencies embodies other subtle meanings. One's feelings may be affected when there is a large change in the exchange rate. People may change their country of residence when there is a long decrease in the exchange rate. Interest rates on loans for building construction may increase so that construction stops. This is analogous to the textile simile where everything is interconnected. Some financiers watch these kinds of subtle changes to decode the underlying conditions of a country.

There are hundreds of indicators of economic conditions. Lanchester explained a few of the more recent indicators in Money Talks: Learning the language of finance. (Lanchester, 2014a) The financial news organizations hint at the relationship of security markets with other aspects of human group life. How security prices affect employment, and affect the cost of living. However, these invisible changes in price influence much of human group behavior. For example, the central bank may decide to intervene in the currency. This may weaken the loyalty for the country; a part of the textile simile.

There are consequences that a financier can estimate. He can speculate on the results. Most of the interconnections are non-linear which means three or more independent variables are influencing each other and all the other variables. He has a model in his mind about how the non-linear events are interrelated. He can anticipate the results and earn money because he knows how to decode all the indicators. When a large number of financiers are making the same move, this becomes an overwhelming force on measures of money such as interest rates. This is similar to how the mind works in fast changing conditions such as driving a car that is out of control. The many factors in the environment influence each other in unpredictable non-linear relationships. And to make the influences less certain, they are more probable than causal. There are countless, invisible levels of interactions between the financial indicators. Rickards used the analogy of the hollow Russian matryoshka dolls which fit exactly into each other. Removing the head of the biggest reveals another doll and so on for endless dolls. (Rickards, 2014) He demonstrated that much of the world of finance is impossible to understand, much less to manipulate toward a goal such as higher employment rate. When the banks manipulate financial systems, the result is often the opposite of the intention because people cannot understand global finance.

When a financier is faced with these enormous forces every hour, how does he explain it to himself so he can make some order from the chaos? Over a long duration of time, he learns some of the non-linear relationships. He learns how certain politicians or other powerful people affect macro economic events. He learns how the relationships are signaled through numbers and measures. He knows the channels through which the communications take place. So bit by bit he can read what depends on certain prices and how those prices change. Changes in price mean different things for each type of financial security. He formulates a hypothesis of the model. He looks for confirmation or falsification of his model. Often he cannot decode the myriad of messages. He may know someone who understands certain price changes so he asks about them. In any case, he knows changes are a signal and that somebody knows what the signal means. He uses other communication channels such as the phone or the email. He uses a computer tool.

Another dimension he needs to know in order to decode the communications is the politics, the varying degrees of power within many groups of people, the office, the country, and the parts of government. Everyone has a distinct but changing place in the grid of the influential powerful people. Other dimensions are the human nature, the psychology, the limits on rational behavior, the emotion driven actions. These affect the bidding and asking prices of securities.

What will inspire action by other people? Consider the analogy of a beauty contest. There are the women and there are the judge's opinions of the women and there are the guesses about what the other judges think about the women. A judge may pick a certain woman to win because he thinks the other judges will pick her. Analogously, the financier talks to other financiers to discover what action the others will take. Are they greedy or fearful?

There are trends which are increasing and decreasing. There are the news organizations which project lies and propaganda. But some of the communications are distinctly true. They are buried in the overwhelming set of orderly and confused signals.

Based on the swamp of chaotic communications, the financier has to hypothesize a means of making money and test the hypothesis. He has to construct many different hypotheses from minute to minute and test them all. He has to manage the risk so that if he loses money it is a small amount. His mind is operating on several levels.

Perhaps, his model is that the underlying reality is communicating the sum of all market forces, the underlying reality. The communications are neither clear nor consistent. What can be discovered about the underlying reality from the many levels of signals? He could conclude that there are basic drives in human groups, feelings of greed and fear, the need to achieve the goal for which the money is being earned, and a major lust for power. He could conclude that power may manifest in government people, in corporate people, in wealthy individual owners of assets, in the mass of common men, or in other sources. He could conclude that men created the interconnections of legal and ethical foundations of power; government laws, and evil intentions. Because there are these foundations, he could conclude that they are necessary because people are lying, cheating and stealing. He decodes all these communications to be convinced of the price of a security the next day.

There is a type of mathematics and several mental systems to evaluate a security

A assuming the reader knows the scientific method it will not be summarized herein. Is there a scientific approach to researching the communications from the underlying reality of the financial world? Is the scientific method capable in any way to discovering the underlying influences in the myriad markets? Yes, hundreds of approaches. Some are mentioned. (Hull, 1989) *Options Futures and other Derivative Securities*, (Chernoff and Moses, 1959) *Elementary Decision Theory*, (Graham and Dodd, 1934) *Security Analysis*.

A human curiosity, perhaps an instinct or an archetype, is gaming. There are games on computers that have similarities to the computer tools used by the financial instrument traders. Many assets are controlled by computers such as the railroad grid, the electric power grid, the airplane flight grid, and the building security grids. They are all monitored on visual screens and are manipulated with peoples' hands. They are all tremendously complex mental inventions. But they have enormous consequences on peoples' lives and on the uses of physical resources. Virtually no one realizes the extent of these World3 inventions and the financial securities games. There are many games and systems conceived and created by people with the intent to get money out of the financial industry for individual use.

There are other systems such as the ocean and the forest not made by people. They also affect the values of financial entities. One must extrapolate these to realize the interconnectedness of everything. (Lewis, 2009)

How to employ the problems in the financial analogy as tools

Hypothesis: the original message from the underlying reality has been encoded and then transmitted into the material world. This hypothesis needs to be tested although it must be true in the financial industry. Therefore, one could employ the untested hypothesis.

Shannon opened the door of communication mathematics. Using the math of communication, one can extrapolate backward to identify the original message before they was encoded. This must be true in the financial industry. (Shannon, 1948, p. 379-423 and p. 623-656)

Consider that the underlying reality expresses a message, $f(t)$ as a function of time, t. Part of the encoding is the transformation $f(t) \rightarrow f(t)$ $\sin(at+b)$ a wave form.

The wave form, $\sin(at+b)$ is an extra message called the carrier. It adds nothing to the rate per unit time in the communication channel. **Hypothesis: The underlying reality communicates many messages simultaneously in a time series in statistical equilibrium.** The total of a given time series is a single function, $f(t)$. There are many of these time series with well developed probability distributions not altered by the change of time, t into time, t+t1.

Let Tq be a transformation group of operators which change $f(t)$ to $f(t+t1)$. The transformation leaves the probability distribution invariant. Wiener gave the math of this transformation 50 years ago. Since then, the signal processing and the enlightened understanding of communications have evolved at an accelerated pace. Consider the following exposition as a brief reminder of a method.

Consider one of these ensembles of the many messages in the time series in statistical equilibrium. Let $f(t)$ be true for the entire history, up to the present, of a time series in a certain class or set. Let the past of only one time series of the class be known. This is sufficient known history to compute, with probable error zero, the entire set of statistical parameters of an ensemble in statistical equilibrium to which that time series belongs. This is true for the case when there is a single varying quantity.

Start with the premise that all physical things are initially born, go through a process and die or disappear. The time series that is known could be a distribution of probabilities for a single variable such as the duration of time, tdeath after birth that an entity dies. Let the time series of the objective ensemble obey the laws of physics because these laws have been stable forever. Thus, one knows the history for all time. (Wiener, 1961, p. 67-69)

This is true for a single time series with one varying variable. It is also true for multiple time series in which there are several quantities varying simultaneously. Wiener provided this math in adequate detail to apply it to a specific investigation. (Wiener, 1961, p. 68-92) Thus, the accelerated mathematical approach which has evolved since Weiner began his research is an aid to decoding the underlying reality.

The analogy of spying on electronic communications

There is a subset of spies who face an unknown environment consisting of electronic signals invisible to the senses. The adversary has purposefully encoded the message and has transmitted it into the channel of electromagnetic radio waves. There are endless methods of encoding and almost as many methods of decoding the signal. A respect for the complex but possible decoding of electronic communications can be gained from Smith. (Smith, 2000)

Decoding is composed of layers of methods such as the following.

Knowing the adversary knows the spy is collecting information.

Knowing the many channels through which the adversary communicates.

Knowing the time of day or week when the signal is being sent so receiving the signal is fruitful.

Knowing there is a hierarchy of messages of increasing importance.

Knowing how much the adversary knows about the decoding methods which are being used by the spies.

Developing new methods of decoding.

Using the theory of probability to increase accuracy.

Persisting without pause.

Developing theories about which encoding method is being used.

Filling in the unknown gaps in the decoded message.

Using all available resources.

These elements are also used in the scientific method of research and discovery.

These principles must be applied to decoding the underlying reality.

Proposal to apply the method of spying on radio transmissions to intercept communications from the underlying reality

On could extrapolate, in an oblique way, the method used to intercept all the nuances of a radio signal. The complete method includes the following.

Interception Phase

1.) Continuous listening thus not missing brief or weak signals at random times.
2.) Continuous listening at several places distant from each other allows discovery of direction from which signal was transmitted.
3.) Listening at all possible wavelengths and to other non-wave based signals with characteristics suspected to be relevant. Non-wave based signals include the natural laws, estimates of new approaches, and intuitive hunches.
4.) Listening close enough to the source to receive adequate signals.
5.) Listening with sufficiently accurate recording to allow repeated analysis by many different people.
6.) Employing humans with superior training, with excellent aptitude, and with wide awake attention to details.
7. Focus on most important interceptions first.

Deciphering Phase

After locating a communication from the underlying reality and recording it, then process it for information.

1.) Subtract noise.
2.) Find a way to filter out separate wave based communications from the vast quantity of all non-wave based communications.
3.) Discover all possible channels of communications.
4.) Find a way to discover all the non-wave based communications from all other sources. For example: Psychic field channel yielding ESP information and extremely low frequency, one shot communications.
5.) Identify analogies to "call signs:" communications for specific receivers from specific transmitters.
6.) Identify individual sources of transmissions.

Intelligence Phase

After processing the communication to yield purified data, convert it into useful information.

1.) Understand meaning.
2.) Embed any deciphered message into network of all messages to get a big picture.
3.) Deliver deciphered information quickly to people who can make the best use of them.
4.) Deliver the associated noise and environment of interception for interpretation of the entire environment.
   (Calvocoressi, 1980)

A brief summary has been proposed to achieve the goals:

1.) Enacting the technical approach to identifying the communication channel from the underlying reality and
2.) Gaining the information transmitted by the underlying reality.

This chapter is merely a mention of the elements of the major effort required to achieve the goals.

If a group of researchers use this approach to guide their search, they will make accelerated progress in achieving the goals. The research has been going for thousands of years. Thus we know it is a difficult research which requires a group of people operating at a higher level of mind and consciousness as outlined in Appendix F.

A suggested system of people emerging to decode the communications that is totally different

The Group of people who decode the signals from the underlying reality must perform functions to ensure smooth operations; functions such as organizing, planning, staffing, controlling and directing. These functions must operate in a superior way.

I suggest that the organism called siphonophorae is a good model because they exhibit emerging and flexible functions by entraining several individual animals. Siphonophores are of special scientific interest because they are composed of medusoid and polypoid zooids that are morphologically and functionally specialized. Each zooid is an individual animal, but their integration with each other is so strong, the colony attains the character of one large organism. Indeed, most of the zooids are so specialized that they lack the ability to survive on their own. This seems like a way for strong and brilliant researchers to work without serious conflict.

Chapter 8

Decoding the many channels of communication from the underlying reality requires a higher level of mind and consciousness of individuals and also of the group of researchers

Many different views of reality and the various approaches to decode the underlying reality have been presented. Can one person be convinced that he has correctly decoded communications to yield reliable information? If the information is reliable, it indicates the characteristics of the under lying reality. Then one person could proceed to use the methods recommended in this book to work backwards to discover the attributes of the underlying reality and to discover the information transmitted before it was encoded. This un-encoded information is another level of the underlying reality.

Based on this conviction, and other decoded information not mentioned herein, one can ask the question again, "What was the original information, the original essence, the underlying principles that were then encoded by the underlying reality before it was transmitted thru the communications channel into the universe?" Are the laws and principles that humans have discovered the final decoding or are there other levels perhaps impossible for humans to discover?

Specific approaches are presented below to decode wave based communications. The problems are to locate where the communications take place and how to intercept them. Solutions to these problems have yielded many layers of underlying reality.

The analogy of higher states of mind receiving communication directly without a sense experience: extra sensory perception (ESP)

The simple mental tools above use analogy as an aid in understanding the underlying reality. One can elevate the mind to use the analogy tools. Then one can elevate the mind to investigate the mind itself. The opposite mental tool can be acquired by letting fall away the process of

Abraham Maslow

Hominid and insects are. The middle is ...
and up to a thousand times as big as a human being is ...
scales. Yet, our tendencies of about look and ...
our data when viewed from the standard ... seem to blind ...
the knew in. (Bod, in, 50BC) See Appendix 11 for a comp... ...
some of the Separate Levels of Mind and Consciousness.

At higher metal levels, we like to reason through levels of
... (Bod, in, 50BC) ... ...

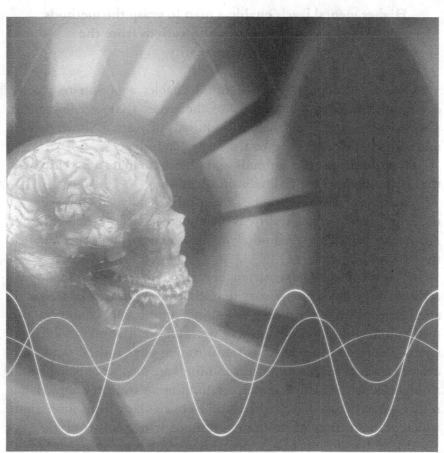

is as follows. The method applies to an individual who is led by
a reader. The individual removes the hindrance to reaching a higher
level of mind. The individual then slowly answers to operate around a
higher levels. Several level of mind is reached by three... that been
liberated from the gravitation of the... phase right down the right even
thinking, meaning, and finally they can bind... ... ... The...
... mental tools to realize whole structures. The ... ... central

thinking and investigating. The mind is then liberated from all visions and habits of thought. In such a liberated state, what does the mind realize? Without distractions of the mind itself and the distraction of sense data, what is received from the underlying reality? Buddha gave the answers. (Buddha, 500BC) See Appendix D: An Attempt to Label some of the Separate Levels of Mind and Consciousness

Higher mental levels enable a man to grasp the methods of intercepting the communications from the underlying reality

One of the higher levels of mind enables a person to receive communications without the use of the senses (ESP). A few gifted people have used ESP in the past hundreds of years. ESP is not well understood. It is suppressed by the science industry.

Hypothesis: people, animals and plants emanate a psychic field that is the channel for ESP communication.

Would experimental proof of the psychic field yield knowledge of the complete underlying reality? Is the underlying reality different from the psychic field? Does the underlying reality create the properties such that the field can exist?

Many spiritual leaders have taught methods to realize the underlying reality. The methods preliminary to realization are usually labeled with another name such as avoiding evil doing, purification, concentration, meditation, absorption, selflessness, devotion, realization of God, and so on. Realization is defined as transforming one's mental states into evolved levels of mind. Some human minds are capable of being raised to higher levels in which some communication from the underlying reality can be decoded. Even then, human beings have many limitations which obstruct complete realization of underlying reality.

Briefly, transforming one's mental states into evolved levels of mind is as follows. The method applies to a single individual who is led by a teacher. The individual removes the hindrances to realizing a higher level of mind. The individual develops his skills to open his mind to higher levels. Several levels of mind are realized in stages until he is liberated from the unwholesome states of mind and from the habits of thinking, speaking and acting that result in pain and suffering. Then he uses mental tools to realize wholesome states. There are states of mind and consciousness of the underlying reality that cannot be experienced without this transformation.

When the individual realizes certain higher levels of mind, he has another method for decoding the principles of the underlying reality. The methods enabled by transforming the mind into a higher state are not used in the scientific methods of decoding underlying reality. A skilled person can decode the entire part of the underlying reality that can be envisioned by the limited faculties of the human being. (Howard, 2012a, Chapter 9)

Hypothesis: a large number of communications from underlying reality are currently decoded.

Then the ensemble of all the decoded properties of the underlying reality, instincts, physical laws, principles enabling life, and so on, could be considered as one massive communication of the encoded principles of the underlying reality.

The effort to understand the underlying reality is formally pursued in a religious context to find out the will of God. In a scientific context, it is pursued to find out the laws and description of the material world. Many themes of the content of underlying reality are commonly invented and used. These themes are often false but seldom tested for causal or probable contradictions of reality. The scientific themes can be considered to be the substrate of the environment, such as atoms, radio waves, micro-organisms and so on.

Much of the underlying reality is not necessarily available to human senses. One could conclude that the underlying reality is an ineluctable property of the environment.

Assume there is an underlying reality that communicates formations of things, changes of things, and termination to all things including the entire environment.

Assume underlying reality is the substrate from which the environment is composed, including the manifest universe and the underlying reality. This is the underlying reality communicating information within itself. The communication is the foundation of all that exists.

Assume that each human invents a mental model of the environment based on all human sense data and ESP.

Assume the underlying reality as modeled by humans, U, can be reduced to the communication process from the underlying reality, C, and the universe substrate, S, that is the manifested underlying reality.

U=C [transformation operation] S where the transform is unknown. The U, including the transform, has always existed as U. The transform is the universe, S, constantly coming into existence, changing in all respects, and disappearing.

Then the themes of the environment invented by religious and scientific processes are more accurate approximations of the underlying reality phenomenon than the sense data transmitted to the brain alone. The themes may indicate the nature of the underlying reality and they may include meaningless misleading noises.

When investigating communications from the underlying reality, the noise can be minimized and the effort can be clarified by starting with the known substrates of the environment. A few humans, mostly higher level minds embodied in spiritual states and in scientists, have investigated patterns of their experience and joined this with the experiences recorded by other humans. This could be labeled 'decoding the underlying nature of reality' but the exact term is 'metaphysics.' It seeks the answer to, "What are the extreme truths of the world?"

The Ancient Chinese believed that there were spirits with which one could communicate

One of the five classic textual recorded foundations of thought in China, was the *Book of Changes* (*I Ching*) which originated before 200BC. It was based on the belief that one could communicate with spirits and with the Tao. Other beliefs were

1.) That the alternation between yin and yang was the Tao.
2.) That which completes Tao is nature.
3.) That which is above determinate form is Tao.
4.) That which is within the realm of determinate form is a concrete entity.
5.) The underlying reality produces two symbolic forms: a broken line and an unbroken line.
6.) Fortunate and unfortunate outcomes produce the need for great actions.
7.) The system embodied in the *I Ching* is without thought and acts (wu wei). It is still and unmoving. That which cannot

be fathomed is the numinous divine which looms behind all change, a mysterious dimension of reality.

These beliefs reveal a powerful influence of the Taoist vision of a world emerging from the Tao.

Discrimination between symbols presented in the *I Ching* and the interpretation of underlying reality received from the *I Ching* as a channel of communication

The *I Ching* was part of a process of communication with the spirits which are called the underlying reality in this book. The *I Ching* or processes like it existed for thousands of years, It was consulted as an oracle, a being who knows more than a human can ever know. The *I Ching* presents various types of symbols. It also presents interpretations of the underlying reality as part of the process of communicating with the Tao, with Heaven and with spirits within various things. One must discriminate between the symbols, the interpretations, Tao, Heaven, and the spirits.

The Chinese living in a higher level of mind clearly discriminated between the experience of reading symbols in the *I Ching* and the underlying reality which was believed to be incomprehensible. That which was written cannot fully explain the meaning of speech. Speech cannot exhaust the meanings of ideas. It may be true that the order and the chaos of Tao cannot be measured or written. If one is confined to words, then one must explain one's intention. So, one must examine oneself. This requires a second level of mind created by the ego. The second mind transmits through the channel of consciousness to the first mind which uses willpower to speak.

A third mind is created by ego. It asks, "What will people think of me?" This mind is examining the first mind, second mind, the body, the memory, the senses, and the feelings. This mind is part of the group mind communicating with other people. This is a repeating category of human experience: examining one's doubts, and one's feelings.

The Chinese sages living in the higher levels of mind established symbols, emblems, and figures to give expression to their ideas. Such symbols, emblems and figures represent realities beyond language. Is it possible to discern these realities if one rises to the higher levels of mind and receives information by means of extra sensory perception (ESP)?

The sages who represented the *I Ching* lived in conformity with Tao. They comprehended principles on which situations are caused. They knew the correct moral response. They gave useful information about action in a tangled circumstance. The sages were the oracle.

The concepts in the *I Ching* never lose their connection to changing situations which recur in human life.

How does throwing coins or yarrow stalks and reading an ancient book relate to an immediate decision about action? At the time a man throws the coins, he is preoccupied by his situational pattern of his existence which includes throwing the coins and believing his answer will come from the oracle. The coins fall in accordance with the laws of probability. He has no way of knowing the best decision to make or the actions to take. He will never have a way of knowing whether his action will be the best path. He cannot experiment by making the situation occur several times. He must believe the *I Ching* will guide him into the best action.

The *I Ching* describes human experiences which are repeated with variations by most people. Thus, men individually and in groups emanated a field of repeating experiences with variations. Perhaps the selected text and symbols resonated with the field produced by the Tao, the ESP, and by the field of the emerging situations. Perhaps the spirits were correlated to the world of Man. Perhaps the mental field, produced by a man's mind which was correlated with the shifting situation, asked the correct question which the *I Ching* oracle answered. The man who asks the question interprets the answering oracle according to his mental preoccupation, his current reality.

The psychologist, Jung called it "synchronicity." (Jung, 1955) (Jung, 1969) Roughly, Jung stated that in the absence of a religious or spiritual expression, such as a church and a priest, the human mind invents connections which exist only in the mind. These mental associations have great power over a man's thoughts, speech and actions.

One could assume that these patterns are the decoded communication signals transmitted into the fragment of reality that is within the limited realization of humans.

Just as the words of the *I Ching* oracle were accepted as truth, other mental inventions have power over a man's thoughts, speech and actions. All large groups of humans have a pattern of inventing a Higher Power, an underlying reality, and then of decoding communications from underlying reality.

Archetypes of the human non-conscious and behavior

The archetype pattern may be received from the communications of underlying reality. In any case, all groups have postulated a Higher Power which provides some explanation for the otherwise chaotic environment. The powers of creating the world, of causing its changes and of destroying parts of the world are attributed to the higher power.

Characteristics commonly attributed to a Higher Power

a.) This structure of thought is labeled, 'religion,' 'science,' or 'spirit.' This pattern manifested as follows.

1. The *I Ching*, its processes, its symbols, its interpretations, and its oracle.
2. The scientifically discovered laws of nature.
3. Religious and spiritual received wisdom.

There exist patterns, archetypes, of human behavior in large groups

Other patterns are called, 'archetypes' of mass human behavior. It may be that group behavior is an expression of underlying reality. Humans have a propensity to practice the following archetypes and instincts.

b.) Archetype: recognition and rigid use of hierarchy in groups. There is also hierarchy in organizing things.

c.) Archetype: the conviction of the reality of one's own ego,
'I am part of something greater,'
'It is part of me,'
'It belongs to me,'
'I am the most important being.'

d.) Archetype: The instinct to war.

e.) Archetype: the instinct to reproduce.

f.) Archetype: the instinct to kill and destroy everything on earth.

g.) Archetype: compete ruthlessly with other humans and all life regardless of the destructive consequences.

h.) Archetype: ignore the law of cooperation but compete, lie, cheat, steal, believe the false, doubt the truth.

j.) Archetype: restlessness, hurry, flurry of wasted busyness.

k.) Archetype: believing in emotion as a voice to obey, such as lust, desire, hate, anger.

l.) Archetype: refusing equanimity, loving kindness, compassion, sympathetic joy as voices to obey.

m.) Archetype: ignore higher levels of mind, ignore the need to examine ones feelings and thoughts.

n.) Archetype: refusing to face the consequences of thinking, speaking and physical acting.

o.) Archetype: ignoring the inclination to do good things.

p.) Archetype: intending to be lazy and slothful, and to do bad things.

q.) Archetype: refusing the instructions to concentrate, to meditate, and to be absorbed in God or a higher power.

r.) Archetype: ignoring the need to increase Chi, to live healthily and simple.

s.) Archetype: ignoring the activation of positive emotions like rapture, optimism, and love.

t.) Archetype: actively denouncing the teachings of the great spiritual leaders which offer the opportunities to activate the potential to rise to the enlightened state which humans are capable of.

u.) Archetype: fear of emptiness: large spaces like desert, void, edge of a cliff, deserted city, black hole in universe.

Universal patterns that indicate they are communications from the underlying reality

v.) All entities are born, live in defined patterns and die. This includes all life on earth and also non-living entities.

w.) Life is a specialized condition which exists only on earth.

x.) When living things cooperate, the group evolves into higher forms.

y.) Great spiritual leaders have taught the decoded communications from the Tao and other decoding systems and have explained them.

z.) Natural language is a gift which can aid cooperation between living entities and in decoding the underlying reality.

aa.) Mathematics is a higher level language for describing the laws of nature and for decoding the communications of the underlying reality.

bb.) Everything changes continuously; non-living things change, even organized living groups evolve. Classes of things evolve into different but similar things.

This decoded information indicates some of the nature of the underlying reality. Based on this information, and other decoded information not mentioned herein, one can ask questions that guide the research.

What was the original information, the original essence, the underlying principles that were then encoded by the underlying reality before it was transmitted thru the communications channel?

How does one identify the channels of communication from the underlying reality?

How to identify the entire set signals and all the forms of the signals transmitted through the channels?

How to identify all the types of signals, the entire taxonomy of subject matter in the signals, and all the classes of information that are never communicated?

Are there laws and principles that humans have discovered that are the final decoding of part of the underlying reality?

Are there communications and included information types that are impossible for humans to discover?

Because the underlying reality is not comprehensible by the limited human mind, then in order to compass, to make real, to cognize, to allow the maximum underlying reality into human experience, an oblique approach is necessary. Such an approach could be described as growing in harmony with it.

Chapter 9

The unsolved problem, an irreducible obscuration, is the fundamental limit on human discovery of the underlying reality

Abstract

The limitation on conscious perception of the underlying reality is partly due to the communication channel consisting of the human senses. Let us overcome some of the limits to the research by proposing hypotheses which are to be verified or falsified. The objective is to propose assumptions and hypotheses of the properties of the underlying reality. This will aid in designing experiments to test whether such properties are indeed correctly identified. Religions and science have sought to discover and describe the underlying reality. The method of gaining subtle and gross knowledge is given. The analogy of the rock on the mountain is given. One can overcome some of the limits to the research by proposing hypotheses which are to be verified or falsified.

There is a limitation on conscious perception of the underlying reality partly due to the communication channel consisting of the human senses

Mountcastle gave a short description of the limit of conscious perception of the underlying reality as communicated through the channel of the human senses, the brain and the mind including ESP. Each person lives in a world created by one's senses, brain, and mind but also bounded by these personal attributes. Each person believes he senses the world exactly as he believes it to be in the exact time he is conscious of it. Mountcastle asserted that the world a person creates in his mind is a perceptual illusion, even a delusion. A person's mind and brain are linked to the body and external world by a consciousness channel, millions of fragile sensory nerve fibers, and ESP which transmit a poor imitation of the reality outside the body, brain and mind. They do not transmit some sensations; they inhibit others. A person can only

be conscious of body parts that he is immediately aware of. Therefore, he is not able to sense most of the body and the external world. Perception of sensations is an abstraction created by the nervous system, brain, and mind. (Mountcastle, 1975) (Hochberg, 1978)

The process of human perception has been studied by psychologists, physicists, and other scientific disciplines for many years. Although most phenomena can be described in detail using scientific methods, the perception of these same phenomena often does not agree with what is measured. An example is distorted memories of criminal events that may result in a false conviction of an innocent man in a trial. Traffic accidents are often caused by faulty perception. Sometimes a substitute reality is desired such as a movie, virtual reality in a game, or a magic act. A machine may be substituted for a human mechanic but the machine uses a different perception. A neuroscientist may wish to discover the bodily processes on which a human depends for sensing. He is usually surprised by the unexpected discoveries.

Most events cannot be detected by human senses, for example, microscopic events or astronomical events. Perception is altered by education. For example, a foreign language sounds like nonsense. After education in the language, one can easily decode the sounds of the language. Of the many thousands of sensors in the body only a few are actually delivered through the consciousness channel to the brain. Most sensations delivered to the brain are not delivered to conscious awareness in the mind. Not everyone receives the same sense data from an identical phenomenon because of the different range of sensitivities in each person. (Hochberg, 1978, p. 1-5)

This is a milder version of what the Buddha experienced. He declared that what is sensed is emptiness. A human superimposes his mentally and emotionally limited presumptions on the world outside his brain and mind.

The physicists of the quantum mechanical revolution inferred the same emptiness.

Mountcastle observed that projecting from the brain are countless fragile sensory nerve fibers arranged in groups uniquely adapted to sample the energetic states of the world about one: light, heat, force, sound, etc. That is all we ever know if it directly. The rest is, perhaps, logical inference. The communication of electro-chemical impulses are all the brain and mind are supplied. The sum total of all information about the world must be decoded by the brain and mind from these countless signal impulses. These signals are sensory stimuli. (Mountcastle, 1975)

The pathways from sense organ to brain are never direct. Sensory stimuli are transduced at various nerve endings. Neural replicas are dispatched toward the brain and mind; to the gray mantle of the cerebral cortex. There are always synaptic linkages from neuron to neuron. There are several of these relay stations. Each of the relay stations and each neuron have the opportunity to modify the coding of the messages from the sensory receptors. Thus, each of the components of the pathways adds noise and delusion to the message. (Hochberg, 1978)

The sensory signals transmitted through consciousness to the brain and mind are the sum of the original signals transmitted by the senses, the additions and subtractions at the synapses and at the Dorsal horn. As of my recent research, not all other modifiers of the signals as they are transmitted have been identified. The brain and mind continually update the individual person's model of the world. Some of the noise is identified and updated in the decoding of later signals. The brain and mind use them to form continually changing, updated neural maps and memories in the mind. This is the illusion of the external world for a given person, of a person's place within the illusion, of a person's orientation of the body, of events affecting the person. At the level of sensation, one person's images and another person's images are virtually the same. It is possible to identify that several people experience the same sensations by verifying through speech or by responding to the world with the same action.

Part of the model is the individual person's location, the individual orientation in respect to the external world, and the individual ego separated from other people. It may be possible to experimentally identify and correlate two people's mental images of a visual object. Beyond that, each visual image is joined with an individual's world model, mental ability, memories of experiences, inferences, and genetic contributions.

There is noise in the sense data, in the perception, and in the experience of the sense data. A person must remove the noise to comprehend the situation. Estimates of the properties of the underlying reality are mental settings which aid in removing the noise. Then a person begins the search for the properties of the world. A person consciously examines the expression of reality within his senses.

From the complex individual use of the sensory information, each person constructs a higher level of perceptual experience. The visual sensation becomes a private property for each individual. The sum of memories, mentally constructed models, new sensory information, lost memories and discarded previous world models is one higher level perceptual experience which is his personal world view.

A person often realizes the potential for incorrect decoding of the messages coming from outside the body. Scientific observation of the brain reveals that even the simplest stimuli are signaled to the appropriate primary receiving area of the cerebral cortex in the form a code. The code is a set of nerve impulses in varying time sequences and in many neurons in parallel channels of communication. That is all that is provided to the brain and mind. Each person's world is mentally constructed out of whatever can be decoded from the neural channels and whatever information remains after the noise is subtracted. (Mountcastle, 1975)

These concepts were later researched experimentally by Smith and others. (Smith *et al*, 2000)

The next problem is to propose hypotheses of the properties of the underlying reality to aid in designing an experiment to test whether such properties are indeed correctly identified

Appendix A lists the assumptions and the hypotheses that are proposed to reduce the problems in this research. In Appendix A. Assumptions and Hypotheses are Introduced to Reduce the Problem of Intercepting the Communications from the Underlying Reality and the Problem of Decoding the Information Received.

Hypothesis: the underlying reality is not dependent on a cause or a probable influence.

Hypothesis: the expression and manifestation in the world of the underlying reality is the result of encoded communication.

Hypothesis: the brain and mind remove the noise from communication by testing the mental model of the world against neural messages received after the model is imagined.

Hypothesis: the most fundamental properties found in the world, or not yet found, are the encoded communications from the underlying reality.

Mathematically, one must define the boundary conditions in order to test the hypotheses.

There are many fundamental properties: laws of nature, waves, non-linear processes, human archetypes such as war, sinfulness and hierarchy, everything changing, time and space dimensions, many aspects of mathematics, weather, the attributes of living beings, the effects of large masses, and reproduction of a species, etc. The list could include hundreds of properties. Estimates of the properties of the underlying reality are mental settings from which to begin the search for verified properties and expressions. If possible, one describes the properties mathematically.

Part of this research is removing the noise and the impossible properties. The researcher identifies the reasonable conditions at the boundaries of a property. This is the beginning point for the mathematical description. See Appendix B Some Methods for Removing Noise from the Communications from the Underlying Reality.

Hypothesis: all identifiable properties are the manifestation of the underlying reality that is being communicated into the world.

Hypothesis: After a person, place, thing, entity, theory or idea has been analyzed to the most fundamental level of the dependent origin, such fundamental level is the underlying reality or an encoded communication from the underlying reality

Hypothesis: the most appropriate descriptive tool for describing the underlying reality is mathematics, a language consisting of explicitly defined terms and unambiguous operators based on rules of logic.

Hypothesis: there exists limitations on how much can be discovered about the underlying reality. Human knowledge has many limitations including intelligence, senses, tools, time, money, natural laws, and unknown limitations.

The method of gaining subtle and gross knowledge

Experiments must be designed to reveal communications from underlying reality. Defining the boundary conditions is part of designing the experiments.

The objective of the experiments is to gain knowledge. The general approach to gaining knowledge of the information transmitted by the underlying reality is based on correct thinking: concentration, meditation, and absorption. If the Feng Shui practitioner is a Buddhist enlightened Arahant or a Taoist transcendent Lohan, then he will concentrate on the interconnection of all factors, not restricting himself to a correlated pair of variables or a small number of independent variables as a physicist would.

An approach to gaining knowledge is concentration on one object at a time. Thus, an unbroken stream of meditative thoughts would extend to the object. He may become absorbed into the object. Then he would perceive the true nature of the object without the distortions of the un-disciplined mind. For example, he will perceive that the mountain at the end of the valley diverts the wind, the water, the light, and the Chi of the animals nearby. He may discern that this is a stormy location not suitable for a residence but good for a windmill to pump water. He will discover that the process of concentration, meditation, and absorption is a proven method to acquire knowledge of matter, of his profession, of modifications to a habitable space that improve it, and of the human mind.

There are many methods to acquire knowledge. Each method depends on the mental and physical preparation of the seeker. More development of the mind through Buddhist or Taoist training yields more levels of mind and more channels of consciousness delivering information to the various mind levels. For example, fewer addictions, less craving, and less dissipation of Chi yield more mind levels. More physical training, more complete nutrients, more control of the mind yield more levels of mind and associated consciousness.

Analogy of reading as decoding the communication from the underlying reality

One can analyze how one reads as analogy to gaining knowledge about the underlying reality. Reading is a method to decode written material. One could apply the reading process to decoding the communication from the underlying reality. Let us consider a stepwise process that begins by defining the anatomy of the senses, defining the interactions of the body with the mind, and introspecting the mental process of reading. One could define the limits of each of the senses and how the mental constructions modify the actual sense data, the words being read. The delusions caused by the mental processes could be experimented with and comprehended. Finally, experiments could be conducted on the various subjects being read, the emotional state of the reader, the eye movements, and the process of understanding the written material. (Hochberg, 1978)

These are a few of the many aspects of reading. They are approaches to perception. Reading has been studied in detail. A huge literature exists about information on sense data related to the mental process of reading. Reading is a subset of the many perceptual tasks that humans perform.

Consider the minor aspect of reading: glancing. Reading speed varies depending on the reader's intention, the nature of text, and the reader's skill. The process is a series of scans or fixed glances. The reader recognizes the words based on familiarity, attention, and expectation. What sets the limit on recognition of unfamiliar text? There is short term memory and, after a time, long term memory of the text. Time is required to decode the text. The mind has several tools for decoding the text. The reader reviews his memory to aid in decoding what follows later. He combines a series of glances and the mental decoding that has occurred. The information in the text is decoded at last.

This process is similar to decoding a secret message intercepted from the electromagnetic field produced by a wireless transmitter.

Consider how this process can be expanded to decode the communication from the underlying reality.

The analog of the rock on the mountain

Many levels of mind are analogous to a higher mountain from which to see further. A stone rolled down from the higher mountain gains more kinetic energy than a stone rolled down a lower mountain. The possible energy in the rock is called "potential." The higher mind has the potential power to deliver greater dimensions of perception to the intellect. Then the many possible higher minds will comprehend chaotic situations, non-linear math, and fluctuations of all relevant influences. In Feng Shui, the result may be an accurate recommendation for the use of a space or a building.

A practitioner who is also an Arahant or a Lohan would design the tools, would write the books, and would make accurate measurements. He would not be limited to the dogmatic notions of the uneducated, or the superstitions of the society, or the fixations on family relations, or the deference to power players, or the demands of overbearing negative personalities, or the impotent laws. He would not allow these limitations to suppress the value of improving the environment.

Mental experiments yield emptiness of reality

All people grasp "at self." Perhaps animals do also. People do not usually question the innate mental state believed to be the self. Does this accord with reality? This misperception of 'innate' does not mean it accords with how things are. Grasping at self is believed to be a form of ignorance. (Dalai Lama, 2012, p. 52)

The scientific revolution was based on the hypothesis that things appear one way but exist in a different way. Scientists accept that there is an underlying reality such as molecules. But most of us do not question that things appear in an independent reality. We accept that our perceived innate self in the world is real and separate. Then we react on the basis of perception. Buddhists and physicists claim that people are not reacting to things as they actually are. We have to investigate the difference between apparent reality and our perception based on noisy sense data about reality.

This scientific or Buddhist point of view differentiates between what an individual person perceives and several levels of underlying reality. The basic foundation of reality is emptiness. Initiating his investigation of this foundation, the Buddha deduced the delusional perception which almost all people believe is reality. Physicists have pursued the same deductions.

The Buddhists and physicists claim the nature of everything is emptiness which is interpreted by humans and other entities based on their thinking, actions, and speech. By undergoing training, eradicating the delusional thinking, and activating the inherent ability of humans to realize emptiness, people can enter an empty state which transcends ordinary reality. The mathematical description of this was written elsewhere in this book.

Buddhists contrast the emptiness of underlying reality with delusion. For a person to be deluded or to hold a strong emotion he must assume a truly existing object. But his mind mentally constructs the object based on his sense data, his previous experience, and his grasping at his conviction of true reality. When his mind is grasping at an object, he has a strong emotional reaction to the object. He is convinced that object is independent of cause; and not caused by his mental construction. However, for a trained Buddhist master who is based in emptiness, none of this is true.

Such a Buddhist has dissolved the delusion. His mind is dependent on his body for existence. Therefore the mind is empty. He no longer grasps at apparent solidity. From the point of view of emptiness, he realizes that the object exists due to the dependent origination; the mind depends on the body for its origination.

For example, if a person hates someone, he imagines many negative traits in the hated person. However, these traits are dependent on mental constructions projected onto the hated one. If a person takes action due to hate, he must face the reaction which will cause his suffering. If his mind awakens to the correct view of emptiness before action, he will realize the emptiness of his hate. Then he will not suffer the consequences of his actions. (Dalai Lama, 2012, p. 86-87)

A psychologist corroborates this view. The seeds of hatred may be sown in a person who has an internal conflict for which the location of an enemy will supply a relief. A sense of personal worthlessness or helplessness characterizes a diminished and desperate individual. Such a person is grasping at the delusion that if he can destroy another person then he will be worthwhile and he will feel relief. (Gaylin, 2003, p. 173)

The opposite of this is to realize the emptiness underlying his feelings of worthlessness, hate, and conflict; the emptiness of hate as an antidote to helplessness.

Various religions and various sciences have divergent hypotheses about underlying reality.

Many scientists in various fields have come to the same conclusion that everything is interconnected and that the human mind arbitrarily separates the parts from the whole.

Hypothesis: the underlying reality is not dependent on a cause or a probable influence.

It is not constructed. It does not rely on something else. If it does not depend on something, it is not possible to describe it in terms of something else or to issue an analogy. (Dalai Lama, 2012, p. 86)

Ancient Chinese Taoists asserted that experiments will not yield knowledge of underlying reality

The Chinese Taoists postulated that everything is one underlying reality which interacts with itself, this constitutes the universe. The Tao influences everything based on unchanging laws and principles. This estimates that the properties of the underlying reality are not possible to discover. This would mean the search will be fruitless. If one assumes that humans will never completely comprehend all the underlying realities, then only parts of the realities can be assumed to be the limit of what is knowable by humans.

Ancient Greeks and Indians decoded underlying reality as countless gods

Thousands of years ago, the Greeks and the Indians invented the pantheon of countless gods. This was based on their life experiments. Consider that these pantheons were their determination of the underlying reality. The Indians and Greeks decided that the universe is composed of many different underlying realities all of which communicate expressions which result in manifestations of themselves in the world. The expressions and manifestations were decoded by various underlying realities to yield the universe. The sum of all the expressions of the gods who communicated their messages to parts of the universe, to parts of itself, is the underlying reality. Parts of the universe, including humans, decoded the messages, such as a typhoon. The decoding process yielded the information about how the gods manifest themselves as the universe including all its physical laws, biological laws, entities, psychic phenomena, animal group behavior and so on.

Hypothesis: the expression and manifestation in the world of the underlying reality is the result of encoded information communicated to the universe as a whole.

These ancient religious concepts and convictions may yield a clue about the properties, communication system, and information transmitted by the underlying reality.

Designing experiments to clarify reality and to decode the communications from underlying reality is extremely difficult.

I recommend experiments to falsify or verify some of the existing mental constructions describing the underlying reality of the universe

The history of the search for the underlying reality began thousands of years ago. Continuing that search is a valuable expenditure of effort and money. One could expect some valuable results from these searches. Cultures based in the mental inventions, World3, of religious systems govern the details f life in whole cities and nations. Curiosity about underlying reality will be satisfied. Perhaps new technology will evolve. Perhaps a new preoccupation will enter the human collective mind to replace war, power addictions, and the destruction of the earth's surface. Perhaps humans will learn to think, act, and speak in harmony with the Earth and the underlying reality.

All creation theories assume there was an entity, God, Tao, emptiness, and so on that existed before the universe became manifest. These theories assume, without basis or proof, that there is a prime mover, such as a God or gods or emptiness, which reacts on something else, the potential universe evolving into reality. This God acting on the universe results in the actual universe. This may be a valid assumption but it is not proven feasible except mathematically and conceptually. It may have value in scientific experiments and mental constructs if it is verified as practical. Experiments are recommended to verify or falsify the creation stories and the theories that describe how the underlying reality interacts with the world.

Chapter 10

Scientific views of the ultimate source of reality

Abstract

There exists the scientific hypothesis: the world as conceived by humans is a poor representation of the underlying reality, its expressions, and its manifestations. All societies have a fundamental tendency to obstruct adaptations of theories or changes in paradigm. Examples of failures to decode the communications of the underlying reality are given.

Hypothesis: The dominating scientific hypothesis is that the world as conceived by humans is a poor representation of the underlying reality, its expressions, and its manifestations.

The Buddha presented evidence that the world is not what we conceive. It is based on emptiness. (Dalai Lama (2012) p. 147 ff)

This hypothesis of poor representation has been verified as a practical and useful theory in many scientific researches in the last 400 years.

Scientific endeavors have found that there is another deeper level that explains most sense data. In physics, the last layer of reality is assumed to be the Higgs Boson, labeled, 'the God Particle.' The Higgs Boson is hypothesized to be the cause of mass. For physicists mass is the end of the search. For the non-physicists, they know that there is more to the universe than mass. On the other hand, there may be no final layer of reality.

The scientific approach has provided the concepts and tools to detect and analyze wave forms. Waves in many different media are observed in most phenomena. Diffusion has been observed in almost everything. Many other concepts have been found in most phenomena such as the physical laws. Living beings appear to obey many biological laws. There has been a search for the principles of law, justice and other relationships between large groups of people or between assemblages of living beings. It is possible that these rules of order are manifestations of the underlying reality.

Humans are partly governed by archetypes and the functions of the dominant half of the brain which is interacting with the self conscious mind function. But human speech and body muscles are also governed by non-conscious activity in the minor half of the brain: the half without any interaction with the self conscious mind function. (Popper and Eccles, 1977, Chapter E5. Global Lesions of the Human Cerebrum, p. 327) Could half of the brain and the self conscious function be channels of communication with the underlying reality? Could the minor half of the brain be a channel of communication with ESP?

Hypothesis: the underlying reality communicates with the world in wave forms and in other forms. Waves, fluctuations, are often observed in physical phenomena; suggesting that underlying reality can be found in wave forms. To communicate, the underlying reality's messages are encoded and transmitted into the existing world and into the possible entities of the world. The communications must be discovered and analyzed. This will reveal the encoded messages. By decoding the messages, some properties of the underlying reality may be identified. This is the central recommendation of this book.

Hypothesis: There exist communications from the underlying reality which affect the world and which are transmitted through an electromagnetic (EM) channel.

Hypothesis: The majority of communications from the underlying reality are not transmitted through an EM channel.

Hypothesis: It is possible that locations, specific times, and known characteristics of the communication system exist where one could find information defined and encoded by the underlying reality

There is the problem to propose where, when and in what form, the communications are taking place. The communications are most probably ubiquitous, continuous, and in many forms. Then experiments can be designed to receive the communications, process the signals, remove noise, and decode them to yield information.

There are common components in all known communication systems. This is shown in Figure 1 above.

Hypothesis: all the known components of communication systems are also present in the communication system used by the underlying reality.

There is another obstacle to the search for underlying reality. It is embodied in governments and cultures.

Fundamental tendencies of all societies to obstruct adaptations of theories or changes in paradigm

In Chinese society as elsewhere, there are tendencies, based on strong convictions and supported by tradition and by powerful people, to retard the adaptation of theories into more valuable scientific tools. (Kuhn, 1970) This applies to arts, sciences, power moves, and cultural traditions.

Among the countless books about Chinese thought and culture, see (Schwartz, 1985). *The World of Thought in Ancient China*. Schwartz discusses Early Cultural Orientation, Chou Thought, Confucius Analects, Mo-Tzu Challenge, Emergence of Common Discourse, Way of Tao, Defense of Confucius, Mencius and Hsüm-tzu, Legalism, Behaviorism, School of Yin and Yang, and more.

He limits his research, "This book will not solve the body mind problem. Or resolve meaning of mental states." (Schwartz, 1985, p. 4)

Examples of failures to decode the communications from the underlying reality

An example of a failure to decode underlying reality was expressed as denial of natural laws of economics, metallurgy, and farming. See Appendix C: Many Different Varieties of Reality describes these real life examples of denying long established realities.

Chapter 11

Underlying reality exists in several layers

The scientific approach formally accepts layers of underlying reality or any other uncertain sources of reality. An objective of scientific endeavors is to uncover the chain of cause and effect. When the causal connection is in doubt, probability mathematics often supplies a handle on the influences that evolve into an event. Another acceptable approach is that some phenomena are unknowable.

Hypothesis: whenever two or more levels of underlying reality are discovered, there is a communication between the levels.

One can study two layers of reality, for example, an aggregated element, such as a copper wire, is one level and the atoms composing it are another level. Thus, the atoms are communicating information to the aggregated element. The information informs the element what properties it has and how to react with other elements. The composite is communicating information to the atoms. The information includes the excess electrons, the oxygen reactions, the stresses and the strains. This type of communication can be studied to reveal the components of the communication process, the encoding process, and so on. Wherever levels of existence are found, they can be processed to uncover the components of the communication system.

The DNA industry uses computer based tools to analyze the details of DNA to decode the communications between levels

Another example is the extremely complex biological molecules such as DNA and medicines which communicate to the ensemble about what reactions are allowable. The ensemble communicates with the molecules about reactions in process, parts of the molecules that are no longer attached, uncertainties in space and time, changes in the angles between atomic bonds, stresses and strains. There are many levels of underlying realities when healing molecules interact with the target micro-organism ensembles. If all the communication systems in

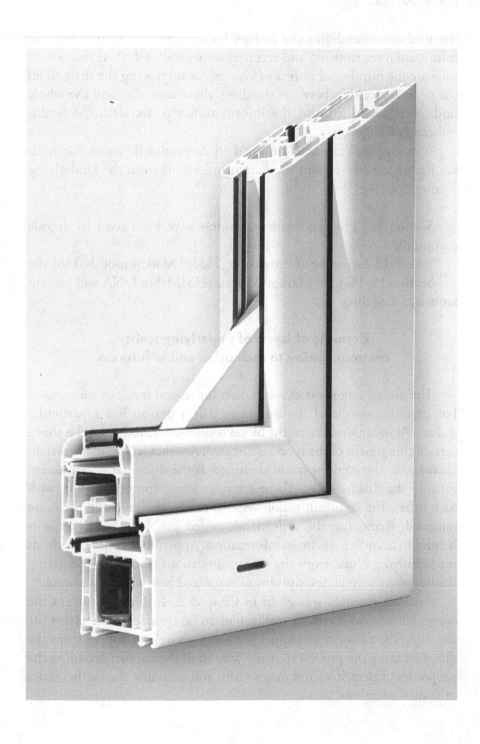

the medication and also the foreign bacteria were identified and the information transmitted and received were cataloged, then this would increase the number of points of view of for improving the drug. If all the communications between the drug, the disease site, and the whole body were identified and the information was decoded, the health industry would evolve.

A recent application is outlined in Appendix B: Some Methods for Removing Noise from the Communications from the Underlying Reality. See

Section 13. Hidden Markov Models have been used to decode communications

Section 14. Decoding of signals using hidden Markov models (HMMs)

Section 15. Hidden Markov Models (HMMs) in DNA and protein sequence modeling

Example of layers of underlying reality communicating to each other and to humans

The underlying realities are known for several levels in some cases. For example, one level is diagnosing that a person has a particular disease. Another level is that the blood is diseased. Another level is that a certain component of the blood is diseased. Another level is that a certain chemical in the component is identified as the discordant component causing the disease. Thus there are four levels communicating with each other. The communication systems can be analyzed and the noise removed. Removing the noise is an added problem. Some noise is helpful in decoding the main information. Appendix B: Some Methods for Removing Noise from the Communications from the Underlying Reality offers several descriptions of noise and how to remove noise.

Another example was given in Chapter 2. In order to invent the telegraph, several layers of reality had to be decoded. The layers are decoding the presence of copper in the dirt, smelting copper from the dirt, inventing the process to make wire from the copper, decoding the properties of electricity and magnetism, and decoding the mathematics of communications.

The human brain and mind operate on several levels simultaneously. This can be experienced as a mind experiment by many people. See Appendix D

An Attempt to Label Some of the Separate Levels of Mind and Consciousness. A mind experiment, introspection, can establish that levels of mind and brain communicate with each other. This is a type of verification of the hypothesis given above: whenever two or more levels of underlying reality are discovered, there is a communication between the levels.

Chapter 12

Conclusions and Recommendations

The justification for the publication of this book is to persuade groups of people to pursue the recommended goals and to accept the conclusions.

Large groups of people have their own collective reality. The group reality has always led to the collapse of the group usually with massive death and destruction. I recommend the expenditure of resources to create a collective reality based on the total capabilities of the human mind and on the total communications from the underlying reality.

Recommendations

Massive resources have been wasted on extending the power, possessions, and prestige of a tiny percentage of elite people in government, business, and religion for the last 3000 years. The major tools used to project this power, possessions, and prestige have been weapons and war. The many tools for expanding the mentally ill egos of the elites are now threatening the existence of life on Earth. This intention to amplify the egos of a few elite people must be diverted into the intention to discover what will aid the evolution of human kind in concert with all life on Earth.

The guidance toward discovery of the correct thinking, speech and actions is as follows. If the collective intention of a group is toward the following list of goals, then it is evolving toward correct thinking, speech and action.

Whatever heals the Earth and whatever extends all life on earth.
Whatever brings people together under the principles of Respect, Harmony, Tranquility, and Purity.
Whatever aids the evolution of human kind, in concert with all life on Earth, toward the following.
Correct view of reality.
Correct thinking.

Correct speech.
Correct, action.
Correct livelihood.
Correct effort.
Correct mindfulness.
Correct concentration.
Correct equanimity.
Correct liberation from suffering.
Correct enjoyment of pleasures.
Discovery of, total reception of, decoding of, and correct use of all communications from the underlying reality.
Discovery of the total powers of and the correct use of all the powers of liberated minds.

I recommend the diversion of all money and human effort now wasted in weapons, war, threats of war, threats of destruction of parts of the Earth held by large groups of people, threats of destruction of the means of production held by large groups of people. I recommend this money and effort be directed toward the valuable goals listed above.

The first research goal is knowledge of all the capabilities of the human mind, especially extra sensory perception. The intent is to discover the total powers of and the correct use of all the powers of liberated minds.

The main objective of this book is to be the engine of directing mental activity, funding and concentration on the power of the mind and consciousness based on the human body. This concentration will be discovered to be orders of magnitude more powerful than military approaches, much more effective in providing the necessary goods and services for large groups of people than totalitarian government approaches, and efficient in allocating human resources. Human resources must be allocated toward creating a world suitable human survival, for positive human expression, for human freedom, and for the survival of the Earth itself.

After making the full powers of the human mind available, I recommend the next research goal is knowledge of the communication with the underlying reality. The intent is to discover the total reception of, decoding of, and correct use of all communications from the underlying reality.

Many suggestions are mentioned in this book for locating, intercepting, observing, decoding and communicating with the underlying reality, and profitably using the information. These suggestions are only a small part of all possible channels and communications from the underlying reality.

Recommendations for research in decoding the underlying reality based on hypotheses about psychic science

The scientific establishment has banned research into the faculty of the mind called extra sensory perception. There is a particular need to test the following hypotheses:

Hypothesis: There are means of psychic communication which do not have the attribute of mass.

Hypothesis: The psychic field is associated with another field analogous to the way the electric field is associated with the magnetic field.

Hypothesis: The assumptions and boundary conditions in the math of chaos may be decoded messages emanated by the underlying reality.

Recommendations for testing these hypotheses are given below. (Bird, 2003)

Libet suggested an experiment to discover the connections between brain-consciousness-mind, "A Testable Field Theory of Mind-Brain Interaction." (Libet, 1994) This should be pursued.

Pandarakalam framed a taxonomy of apparitional experiences, the major observed events, and the supporting medical and psychic observations. This can be used as a starting point to experiment with psychic expressions of mind. The taxonomy can be the initial structure for experimental designs to verify or falsify the hypotheses and assumptions suggested in Appendix A below. (Pandarakalam, 2011 and 2012)

I recommend research in training a human receiver or inventing an inorganic receiver. In addition, I recommend research in psychic science to provide many time series of decoded communications from the underlying reality. I recommend that the methods initiated by Shannon and Weiner, which resulted in an enormous body of mathematical methods and electronic decoding equipment, be brought to bear on the encoded communications. The electronic methods used for electromagnetic waves can be extrapolated by analogy to other channels of communication. I recommend that each type of communication from the underlying reality be represented mathematically and decoded. I recommend that the sum of all the channels of decoded communications could be collected to characterize the underlying reality.

I recommend that a small fraction of the effort put into weapons, energy, particles, war and waves be diverted immediately to yield a vast amount of wisdom about the human mind, and consciousness. See Appendix F: Decoding Several Levels of Reality in the Moral Mazes of Thinking, Speaking, and Acting by Chiefs in Government, Corporations, and Religions

Simultaneous to that yield of wisdom, the total nature of the underlying reality could be discovered. If focused on the underlying reality, the concentration, meditation, and absorption of the mental efforts of thousands of mental giants on weapons, energy, particles and waves would have produced astounding results. But they have instead produced vast destruction of the earth.

I recommend that some of these mental efforts be directed immediately toward training people to purify their minds so they can discover the total capabilities of the human mind. Then they can record the many communications from the underlying reality, decode them and apply them directly in the evolution of human kind. A stepwise total diversion from government and war to research into a focus on the total capabilities of the human mind and the underlying reality is required.

Human kind has reached the crossroads when a choice must be made: either destroy the Earth or expend efforts on improving the psychic conditions, the minds, and the consciousnesses of humans and other living beings. Consider how great would be the gain from the concentration, meditation, and absorption of the mental efforts of thousands of mental giants studying mind-brain and consciousness.

I recommend stopping the efforts of a myriad of intelligent people who serve those holding major power over humankind. Those holding power: politicians, military leaders, religious leaders, and corporate leaders tend to cause destruction of all life forms and tend to waste the resources of the Earth.

Governments have always intended to have total control of people. The result of thousands of years of government has triggered chaos among large groups of humans and among the natural processes of the Earth itself. People must stop serving these destructive interests. People must select which major group they will work with, not based on criteria of money and power.

The efforts of a myriad of intelligent people are now being wasted in the service of governments, destructive corporations and hypocritical religions. Their efforts are perverted by acting on the intention to help the cunning and perverse people in power to acquire more power, to control a world full of humans, to gain money and corral material wealth. It is obvious that people with this lust for power, possessions, and prestige beyond any useful magnitude are mentally disordered.

I recommend the efforts of those assisting the destruction of Earth, the extinction of all life, and the degradation of the human species be redirected to discover the communications from the underlying reality and to discover the full capabilities of the human mind, brain and consciousness.

I recommend that the positive ethics and morals of living beings activate human efforts toward conserving spiritual achievements revealed to human kind by the great spiritual leaders.

Recommendation: Remove the obstruction against discovering the underlying reality and the full capabilities of the mind, brain and consciousness

The avoidance of studies relating the human brain, mind and consciousness by the scientific industry is obvious. These topics have been unacceptable for study by most scientists except for some serious researchers in philosophy, psychiatry, and psychology. I recommend that research in brain-consciousness-mind in relation to the human brain and body be vigorously pursued. This pursuit will yield infinitely more than the dedication of such science for war and for accumulation and projection of more power, ego gratification etc. for the already powerful people.

For example, a benefit would be that the enormous mental health industry and the physical health industry would be able to deliver more effective service. Another benefit would be less mental disorders and the associated suffering. Other benefits would be those preached by the saints and sages throughout all of history but never taken seriously by virtually all of human kind.

When the obstruction to mind, and extra sensory perception (ESP) research is removed, one could describe the scientific endeavors as an effort to decode the underlying reality. The psychic phenomena such as ESP have not been extensively decoded.

A central obstacle to using psychic science to decode the underlying reality is the lack of acceptable instruments for detecting, recording, and measuring extrasensory messages. For psychic research, I recommend using the obvious instrument: the human mind and consciousness. But, in general, this instrument is limited by endless distractions.

I recommend that human mind and consciousness of scientists be purified by Buddhist practices. Some distractions would be removed from the mind-brains of the scientists. Then scientific experiments could be carried out to verify or falsify the assumptions and hypotheses in Appendix A using human mind and consciousness as a reliable instrument. The mind-brain instrument could be enhanced to receive of the communications from the underlying reality. This assumes that the researchers have purified their minds and consciousness by applying the recommendations at the beginning of this chapter.

The brain-consciousness-mind are almost ignored in education and in almost all sciences. To the contrary, I recommend that the mind and consciousness be investigated by using the massive funding and intelligence now being wasted on weapons, war and on the projection of excessive power by people already in power.

It is obvious that the benefits of fully using the mind, brain, and consciousness by human kind would be of more value than war, more value to the collective human kind than totalitarian control of people, and more value than the destruction of the Earth.

There is no taxonomy of the entire set of properties of non-material mind and consciousness. I recommend producing such a taxonomy.

I recommend projects to evaluate the methods of decoding the communications from the underlying reality. Many methods have been mentioned in this book. The methods should be evaluated for their efficiency in receiving information.

Another approach is to reduce the effort by decoding only the type of message that is embedded in a wave; only search for channels where waves represent the information.

Further reduce the effort by limiting the search to channels carrying electromagnetic wave forms.

No means of information transmission by radio is as efficient as amplitude modulation (AM). One could ask, "Does the underlying reality modulate the amplitude of a message?" This may be a fruitful question to answer. Is it feasible to develop a means of finding AM in the expression of the underlying reality? Many earth events have a frequency and amplitude which allow accurate and efficient decoding.

I recommend seeking an analogy to the AM used in radio in the efficient analysis of communications from the underlying reality. The final objective of this is the reception of the AM communication by human senses or by scientific instruments.

All the proposed assumptions and hypotheses listed in Appendix A.

Assumptions and Hypotheses are Introduced to Reduce the Problem of Intercepting the Communications from the Underlying Reality and the Problem of Decoding the Information Received

need to be researched. In particular, I recommend research into the unknown and hidden abilities of the brain, the non-conscious mind and the consciousness connecting them.

This would stimulate activity in the fields of mind, brain, consciousness and cognition. Such research would reveal verifications and falsifications of the assumptions and hypotheses.

The purpose of recommending tests of the following hypothesis is to stimulate experimental designs to verify or falsify it.

Hypothesis: Psychic information is transmitted through particles, waves, or another means. This means of communication has neither mass nor energy dissipation.

The testing, research, and discovery would yield valuable knowledge of the following.

a. Definition of the properties of psychic particles, waves or other channels that carry the communications.
   Channels without mass can move information at the speed of light or faster without violating the premise of the theory of relativity that velocity of mass is limited by light speed.
b. Definition of the effects caused by psychic particles, waves, or other channels.

c. Definition of the probable influences of psychic particles, waves, or other discoveries on the mind, consciousness, human senses, living matter, and non living physical material.

d. Methods to identify noise in the communications from the underlying reality. Also methods to remove the noise.

e. Decoded properties of the psychic particles, waves or other channels of communications based on what is known about them after all spurious premises, noises, and assumptions are removed.

f. A non-physical or physical instrument to detect psychic particles, psychic fields, waves, or other characteristics in the communications from the underlying reality.

For example, certain humans can detect the psychic field. They are the instruments.

g. A means of generating many psychic particles, waves or other channels as desired to allow adequate investigations.

h. Definition of experimental conditions to detect the causal and the probable influences of a psychic particle, wave or other channel, or a stream of these.

j. Methods to test hypotheses by the means given by Libet. (Libet, Summer 1994)

k. Novel means of experimenting.

l. The energy or lack of energy of the psychic particles, waves or other channels.

m. Means to detect and measure the psychic field.

n. Definition of, detection of and measurement of Chi.

o. Means of detecting and measuring the causal effect of the psychic particle, wave, or other channels and the probable influence of these means on physical objects.

The following hypothesis was mentioned above as a candidate for testing.

Hypothesis: The psychic field is associated with another field analogous to the way the electric field is associated with the magnetic field.

This may be the consciousness field or the Chi field.

Testing this hypothesis allows the investigation of psychic phenomena using Maxwell's equations for electromagnetism as a starting point. The research requires identifying the two fields, their causal and probable connections with other human minds, and inventing instruments for receiving the communications, measuring them, and recording them.

I recommend that researchers apply the scientific methods, the mathematics, and the human concentration of mental vitality to the study of mind. Employ concentration of consciousness and how it is associated with the human body. This requires people skilled in meditative imagination who are aware of the various levels of mind and consciousness.

Hypothesis: The researchers need to concentrate, meditate and be completely absorbed in the non-material reality to receive and decode some of the communications from the underlying reality.

I recommend that, after researchers have purified and concentrated their minds in accord with the guidance above, that they then investigate the non material reality of mind. Investigate the quantum theory of wave fluctuations and the uncertainty principle as applied to creating non-material, that which does not have mass, psychic particles, waves or other vehicles for communication with the underlying reality with no mass. Apply the existing and novel experimental methods to defining the psychic field, the Chi field and other factors of mind and consciousness. Use this research to decode the communications from the underlying reality.

I recommend that researchers adopt the objective of using the decoded communications to enlighten people to act in harmony with the principles of the underlying reality.

I recommend projects to evaluate the methods of decoding the underlying reality for the property as efficient information transmitters.

Examine communication of wave forms and other generally ordered characteristics in the environment suggested above.

An excellent return on investment can be expected from research into other hypotheses directed at mind, consciousness and their emergence from the body.

This may shine light on the consciousness field. This may define the Chi field, long pondered by the Chinese.

Conclusions

An obvious conclusion is that an intelligent, educated, and spiritually grounded population would allocate resources for research to discover the total technology of communication with the underlying reality and for research to discover the total capabilities of the mind, especially applications of extra sensory perception.

Another conclusion is that these researches would yield the highest return on investment by any measure.

Another conclusion is that government has been the most destructive enemy of human kind for thousands of years. Therefore, human kind must be brought together based on such principles as respect, tranquility, purity, and harmony, not hierarchy and authority.

This conclusion is clarified in Appendix F. Decoding Several Levels of Reality in the Moral Mazes of Thinking, Speaking, and Acting by Chiefs in Government, Corporations, and Religions

The main payoff from the recommended research, and the achievement of the list of guidance above, would be the appropriate employment of people in pursuit of a human population that supports the recommendations above. Such an environment could lead to groups of people who intend to integrate into their lives the list of correct factors above. Perhaps such a population would add vitality to the Earth instead of destroying it.

Conclusion about conditions that did not yield the promise of beneficial effects

The following clarification is a brief history of the attempts to establish governments that provide the necessities.

People want liberty, less suffering, more pleasure, and freedom of expression. Give educated and spiritually grounded people enough freedom from disabling laws and from the suppression by governments composed of mentally ill people. Then people will have the freedom of imagination to evolve into the most appropriate social systems.

However, there are many levels of leaders and many levels of followers. Each person has his own thoughts and his own ego to satisfy, his own evil and his own fine qualities. Not all people want to do what everyone else does. There are disasters that are impossible to foresee. Some rules and laws seem perfect at first but later consequences prove them to lead to totalitarian solutions. The endless complications of large groups were foreseen by the founders of the United States of America and by the many Chinese who framed government systems. The enormously complex nature of large human groups begins to express itself. Then the government fails.

The complex moral maze of corporation organizations in America is given in detail by Jackall. (Jackall, 1988) These same personal and organizational desires motivate government and religious organizations. Reading Jackall's book, *Moral Mazes*, allows one to realize why governments and religions fail in the same way that corporations fail.

Simple example of failed promises: the discovery of scientific laws promised improved living conditions for large groups. However, the laws were often perverted in many ways that degraded life. A case is the discovery of thermodynamic laws which yielded steam engines. But living near a coal mine is degrading. Another case is the discovery of electric power generated by nuclear particles. However, more than one nuclear electric plant has failed in a catastrophe. As I write, the Pacific Ocean is being killed by a failure of the Fukushima Nuclear plants in Japan.

Another type of situation with great promise was the discovery of DDT insecticide which saved millions of lives from malaria, yellow fever and other diseases. In this case, the catastrophe was caused by not using it. It was found to kill animals and birds in addition to insects. Governments and corporations decided it is better to have millions of people die than to have millions of birds die. DDT was outlawed.

Complex human nature, unforeseen consequences and other reasons have destroyed all governments, business organizations and individual branches of religions

The most difficult problem facing human kind is framing a government or other means that would lead large groups to cooperate in a reasonable way, not involving war. The goal for the group behavior is that all the appropriate goods and services are delivered to all the people for more than 300 years without devolving into totalitarian horror.

Profound conclusion: Preservation of the Earth requires cooperation of most humans all over the Earth

The fundamental conclusion now in the context of human history is the preservation of the Earth. This objective can only be done by reducing the power of governments. A means of persuading all people to cooperate must be framed, but not by setting into place a totalitarian hierarchy that suppresses most humans. People must cooperate by actively participating in achieving the goals set in this chapter by following the guidance listed.

Appendix A: Assumptions and Hypotheses are Introduced to Reduce the Problem of Intercepting the Communications from the Underlying Reality and the Problem of Decoding the Information Received

Introduction

In the evidence of the many years that humans have been trying to detect the underlying reality and to define its attributes, there has been meager success. There are many layers of underlying reality. There are few instruments to aid the limited investigative abilities of humans. Therefore, the assumptions and hypotheses in this appendix are limited to a subset of all existing communications to the world from the underlying reality. The subset is composed of the communications that are possible for humans to intercept.

Appendix B is focused on one aspect of discovering the underlying reality, removing noise from the subset of communications.

The simplified approach is summarized as follows. Certain assumptions reduce the problem. The reduced problem allows a solution to be a member of the set of possible solutions.

Assume the following reductions.

Assumptions

It is possible for humans to intercept and to decode communications from the underlying reality in the form of waves.

Defining the entire underlying reality is too ambitious or impossible.

The underlying reality communicates many messages simultaneously in parallel channels in a time series in statistical equilibrium.

The underlying reality communicates many messages simultaneously in parallel channels; each channel transmits messages in a time series in statistical equilibrium.

A few humans, mostly scientists, have investigated patterns of their experience and joined this with the experiences recorded by other

humans to create the body of scientific knowledge. One could assume that this body is the decoded communication signals transmitted into the fragment of reality that is within the limited realization of humans.

There is a reality that underlies the entire world.

There is a reality underlying the world, matter, processes, and life. that can be sensed by humans.

The underlying reality manifests as the world.

The manifestation of the world is the result of a communication process.

The underlying reality communicates the world, the physical properties, the changes, the processes, and the natural laws to the manifested reality accessible to humans.

The fundamental law of all change is a decoded principle of underlying reality.

The following assumptions reduce the resources necessary for the investigation of the underlying reality.

There is a way to discover the underlying reality in detail.

The mental and physical tools invented to discover reality can indeed be discovered.

The underlying reality can be investigated with the promise of achieving a method of defining it.

The underlying reality discovered by the tools can be tested, experimented with, measured, verified or falsified.

The mathematics of communication accurately describes the communication from the underlying reality to the entire world.

It is possible to succeed in discovering the means of transmitting information from the underlying reality into the manifested world.

It is possible to describe the entire underlying reality for a given manifestation, m. This description may be mathematical, either based on human sense organs, or in the realm beyond what it possible to detect with humans senses.

It is possible to discover how the underlying reality encodes information so it manifests as a feature of the world.

It is possible to decode information found in the world to identify that it is a communication from the underlying reality.

It is possible to identify the communication from the underlying reality; its content or the cause or the probability of expression of the underlying reality or manifestation in the world of the underlying reality.

It is possible to separate the encoded information communicated from the underlying reality from the noise in the communication.

The noise can be measured or defined. And the extent to which it distorts the encoded information transmitted from the underlying reality can be known.

It is possible to decode the communication from the underlying reality to discover the content or cause, or probable influence of the information before it is encoded by the underlying reality.

The limitations on knowing the underlying reality can be known.

It is possible to discover the different kinds or classes or levels of information that are transmitted from the underlying reality.

The properties of the underlying reality can be determined.

The communication from the underlying reality can be known separately from its manifestation in the world.

Assumptions must be framed to aid in the discovery of a reduced set of transmissions that are tractable. One could assume that the patterns, detected and hypothesized to be the decoded communication signals transmitted into the fragment of reality that is within the limited realization of humans, will be tested and confirmed to be such signals.

Some conditions of this research due the limitations on the human investigator

A human creates a model of the environment based on sense data and on certain abilities inherent in the human. All these abilities are limited.

It is doubtful that humans can perfectly comprehend of all elements or reality and the source of underlying reality.

Human abilities are:

Inadequate to receive the total phenomena of reality,

Inadequate to decode the entire signal transmitted by the underlying reality,

Inadequate to create a valid model or total reality due to limited mental faculty,

Inadequate to use the decoded information in the best way due to extremely limited mental abilities, and

Inadequate to respond to the signal received from the source of underlying reality due to limited physical ability.

Since humans are so limited, the human has an abbreviated understanding of the phenomenon being transmitted by the underlying reality. The information encoded in waves and received by the senses does not represent the entire communication.

Humans seeking the underlying reality have inherent limitations on the extent they can discover the communications from the underlying reality as follows:

Human senses are limited to receiving certain types of wave representations of the environment such as sound and light and other representations such as smell, flavor, and physical objects of touch.

Humans can invent and produce limited equipment to intercept and to receive communications from the underlying reality.

The communication system between the underlying reality and the world including the humans consists of many parts each of which could become impossible to characterize adequately.

All parts of the communication system are based on probabilities or cause-effect relations or unknown influences. All have inherent uncertainties which humans may not be able to determine.

The reception of information is part of an information communication process all of which humans may not be capable of discovering.

The following information communication system is composed of several elements which humans may not be able to discover adequately to be useful. A phenomenon exists as the underlying reality or some entity in the environment. The communication is initiated and is encoded as the original phenomenon. The phenomenon is encoded into a signal. The signal is transmitted into the world which may include human senses or reception instruments. Noise is introduced. The signal and noise are transmitted through a channel. The world which may include a human receives the signal and noise. The world which may include a human removes noise and decodes the signal. The world decodes the signal by manifesting the information within the signal. The manifestation of the underlying reality is an element of the world. Or, the human nervous system or electronic receiver delivers the signal to the human mind and body for use. Three results may occur:

a) One or more of these elements may yield false results, or
b) A true decoding may seem false, or
c) A false decoding may seem true.

Assumptions that enable a preliminary identification of the underlying reality as a beginning of the search based on information theory and on wave based phenomena

Assume that a dedicated person can begin a decoding process based on received signals that yields the mind objects, the discovered laws, the identified instincts, and the formulated archetypes.

Assume the symptoms of information transmitted directly from the originating source of underlying reality follow.

1. Distinct patterns, taxonomy, mind objects in classes of things or in all things without exception that are detected through observation.
2. Persistence of the mind objects, inventions of the mind, through all phenomena, or all animals or all things whether living or non-living.

Assume that the mind objects, laws, instincts, archetypes, that are identified as expressions of information originated from the underlying reality, can then be decoded to discover the original message.

The basic problem of locating the communication channel from the underlying reality to the universe

How does one perceive which are the waves that one is interested in? How does one know which waves are going to lead one backward to the source of the emitter of the wave? What is the medium in which one finds the waves? What is one allowing oneself to receive through the information contained in the wave? What does one mistakenly assume is not information from the underling reality? How does one decide to expend resources to decode the information in the given waves? So what kind of originating information is one seeking? How would one verify whether the decoded signal is from the originating underlying reality?

The case when the human psyche is the channel which carries the information

Let us consider focusing on the human psyche as the channel. The receiver in the psyche may be the conscious mind or the non-conscious mind. Neglect the communications that are missed.

Next, the received entities diffuse from the psyche field and the non-conscious psyche into the conscious psyche. (Howard, (2012c)

One could create a taxonomy of all communications for reference.

Is it probable that all scientific laws, all defined instincts, and all archetypes of the psyche could be collected together in a single concept, such as ordering them into a symbol or an image. Would the symbol constitute enough transmissions from the underlying reality to infer the structure or nature of the underlying reality?

The EM radiation signal in a neuron must have an electric and a magnetic radiation correlate. This was researched in depth. The diffusion of information from the underlying reality into the nervous system and into the consciousness mind may be described mathematically. It could be researched and described as a mechanism for the brain to yield the psyche field. Writing the equations of specific neuronal circuits was proven possible by Hodgkin and Huxley. (Hodgkin and Huxley, 1952) This may be helpful in explaining the operation of large parts of the nervous system, not just in the brain. This would yield fruitful research.

Limit the research by proposing hypotheses which are to be verified or falsified. These hypotheses are meant to stimulate the imagination of investigators to design experiments to verify or falsify them.

Hypothesis: The researchers need to concentrate, meditate and be completely absorbed in the non-material reality to receive and decode some of the communications from the underlying reality.

Hypothesis: Psychic information is transmitted through particles, waves, or another means. This means of communication has neither mass nor energy dissipation.

Hypothesis: The psychic field is associated with another field analogous to the way the electric field is associated with the magnetic field.

Hypothesis: The assumptions and boundary conditions in the math of chaos may be decoded messages emanated by the underlying reality.

Hypothesis: whenever two or more levels of underlying reality are discovered, there is a communication between the levels.

Hypothesis: The dominating scientific hypothesis is that the world as conceived by humans is a poor representation of the underlying reality, its expressions, and its manifestations.

Hypothesis: there exists limitations on how much can be discovered about the underlying reality. Human knowledge has many limitations including intelligence, senses, tools, time, money, natural laws, and unknown limitations.

Hypothesis: the brain and mind remove the noise from communication by testing the mental model of the world against neural messages received after the model is imagined.

Hypothesis: people, animals and plants emanate a psychic field that is the channel for ESP communication.

Hypothesis: the original message from the underlying reality has been encoded and then transmitted into the material world.

Hypothesis: the archetypes in the human non-conscious diffuse through consciousness into the mind.

Hypothesis: the mind objects, laws, instincts, archetypes, that are identified as expressions of information originated from the underlying reality, can then be decoded to discover the original message.

Hypothesis: the underlying reality is nothing or does not exist.

Hypothesis: the underlying reality is a collection of all the discovered and undiscovered laws of nature.

Hypothesis: an attribute of the underlying reality is that all the emanations from it are interconnected.

Hypothesis: the underlying reality itself has the property of interconnectedness.

Hypothesis: the archetypes in the human non-conscious mind diffuse through consciousness into the mind.

Hypothesis: waves are part of the decoded communications from the underlying reality.

Hypothesis: the conscious mind is an emergent system based on the entire physical substrate, especially the totality of all information transmissions.

Hypothesis: a waveform is a subset of all possible channels through which the underlying reality transmits communications.

Hypothesis: limiting the set of the communications to those within the assumptions and the hypotheses set forth herein does not distort the knowledge of all existing communications.

Hypothesis: Chi is created by the same mechanism that creates subatomic events and particles

Hypothesis: Psychic information is transmitted through particles without mass.

Hypothesis: the universe is an interconnected entity.

Hypothesis: the variables within the information from the underlying reality include the natural laws, the wave form of communication from the underlying reality to the world, the human mental archetypes, the human instincts, the local creation of mass, energy, length, time, the harmony of the underlying reality with the world, and the chaotic conditions necessary for the emergence of life etc.

Hypothesis: the properties of the underlying reality are the boundary conditions at which one can begin experimental investigations

Hypothesis: a large number of communications from underlying reality are currently decoded.

Hypothesis: the underlying reality exists in emptiness.

Hypothesis: the underlying reality includes a collection of all the discovered laws of nature.

Hypothesis: the underlying reality can not be known by definition, because of some other initial conditions, and the limited and points of view available to humans.

Hypothesis: people, animals and plants emanate a psychic field that is one set of channels through which the underlying reality communicates.

Hypothesis: wave components of the underlying reality are commonly observed but not recognized.

Hypothesis: the properties of the inherent makeup of the underlying reality would manifest themselves within the spider web, spider signals, spider life cycle, and also the earth, man and heaven.

Hypothesis: the underlying reality manifests itself within the observing human, the human senses, human nervous system, human mind, the thinking processes of the mind, and the analogies invented by the mind.

Hypothesis: the brain and mind remove the noise by testing the mental model of the world against neural messages received after the model is imagined.

Hypothesis: the most fundamental properties found in the world, or not yet found, are the encoded communications from the underlying reality.

Hypothesis: all identifiable properties are the manifestation of the underlying reality that are being communicated into the world.

Hypothesis: After a person, place, thing, entity, theory or idea has been analyzed to the most fundamental level of the dependent origin, such fundamental level is a property of either the underlying reality or an encoded communication from the underlying reality.

Hypothesis: the most appropriate descriptive tool for describing the underlying reality is mathematics, a language consisting of explicitly defined terms and unambiguous operators based on rules of logic.

Hypothesis: there exist limitations on how much can be discovered by humans about the underlying reality. Humans have many limitations including intelligence, senses, tools, time, money, natural laws, and unknown limitations.

Hypothesis: the underlying reality is not dependent on a cause or a probable influence or any other thing or process.

Hypothesis: the expression and manifestation in the world of the underlying reality is the result of encoded communication.

Hypothesis: the underlying reality communicates with the world partly in wave forms.

Hypothesis: There exist communications from the underlying reality which affect the world and which are transmitted through an EM channel.

Hypothesis: The majority of communications from the underlying reality are not transmitted through an EM channel.

Hypothesis: It is possible that locations of the communication system exist where one could find information defined and encoded by the underlying reality.

Hypothesis: It is possible to determine the attributes of the underlying reality by decoding the received information from the underlying reality

Hypothesis: all the known components of communication systems are also present in the communication system used by the underlying reality.

Hypothesis: whenever two or more levels of underlying reality are discovered, there is a communication between the levels.

Hypothesis: Each part of an animal or inorganic entity has many cells that receive slightly different information from the underlying reality: various frequencies of EM, delayed reception, dependent on the function of the cells.

Hypothesis: The consciousness and mind are generated by the changing processes of the physical brain and body.

Hypothesis:Reciprocally, many changes in the physical brain and body are caused by the conscious mind, World3.

Hypothesis: The mind is composed of a summation of electromagnetic waves, consciousness waves, and psychic waves.

Hypothesis: There are electromagnetic fields, consciousness fields, and psychic fields. They are self organizing because of their fluctuating interactions.

Hypothesis: One probable model of the universe is an information processing system in which the inputs from the underlying reality are undetermined.

Hypothesis: There is a class of communications from the underlying reality that is ergodic.

Hypothesis: There are stationary time series that continuously manifest as a physical part or a mental part of the world.

Hypothesis: The underlying reality communicates many messages simultaneously in a time series in statistical equilibrium.

Recommended research

Each of the hypotheses and assumptions introduced in Appendix A could yield valuable experimental and mathematical discoveries through a system of testing for veracity. As a minimum, research into each hypothesis and assumption is recommended with the objective of verifying or falsifying all of them.

Appendix B: Some Methods for Removing Noise from the Communications from the Underlying Reality

Introduction

Most of this Appendix B is limited to wave based communications operating under the assumptions and the hypotheses listed in Appendix A. Other types of noise are also considered because there are so many types of noise.

Consider a simple case. People who one only knows online, especially in a dating website, must be viewed with suspicion. A person can show any pictures and claim these are pictures of him. He can tell any lies and hide ugly truth about himself. One wants to believe that the online person is what is represented in pictures and in written words. However, one must reserve belief until a real life meeting and until some real world events are experienced together. Then one can know more parts of the other person whom one has met online. This is the desire to remove noise from the communication.

We do not know people when we first meet them. After several meetings and especially after difficulties, we know them better. This is removing some of the noise. We may never know some people very well no matter how many times we meet them. When people are only known online, they are represented by very little information. The information may be intentionally false. The people have some reality but the lies and deception are noise interfering with knowing the person better. This is an analogy to other communications in the EM channel or in the observation of nature channel.

All explanations of a given person have to start out from certain definite conjectures that one makes. Therefore, all explanations of another person's reality is in some sense intellectually unsatisfactory and limited. They are limited because the conjectures we assume to be true are themselves used as unexplained assumptions for the purpose of explaining the person to oneself; a circular argument. One may become conscious of a need or a wish to explain the conjectures and the assumptions in their turn. But this leads to the same problem again: one assumes some information about the person is true. The assumption is

used to convince oneself that other knowledge about the person is fact. This is circular reasoning. So one has to stop somewhere.

In this way, one arrives at the doctrine of the non-existence of ultimate explanation of a virtual person online. Everything that is very important is unexplained, especially everything in connection with the existence of a virtual person in the real world.

One knows that the world exists and one exists in the world but knowing this about a person online is inexplicable.

This Appendix B mostly applies to wave forms. This is because the mathematics and the electronic equipment are available to make this wave form signal processing possible. The general communication system is represented by Figure B.1 Generalized Communication and Signal Processing System

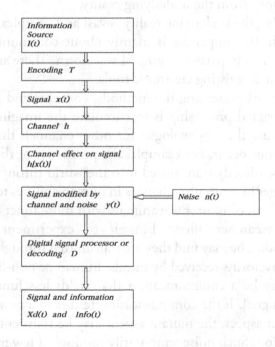

Figure B1 *Generalized Communication and Signal Processing System*

(t) means the function depends on time, t

The following concepts are for the purpose of removing noise from the information within a communication from the underlying reality. They are based on methods and equipment used to remove noise from radio, television, and other electromagnetic (EM) communications. These methods and equipment are restricted to the EM wave forms.

The methods recommended and discussed in this appendix are focused on limited to methods of removing noise from the waves. However other methods of removing noise from other manifestations of underlying reality are also included.

Hypothesis: limiting the set of the communications to those within the proposed set of assumptions and the set of hypotheses does not distort the knowledge of all existing communications.

Whatever is discovered in this research on wave forms will be found not to obscure the discovery of the entire collection of all possible communications from the underlying reality.

Discovering the underlying reality is vast and complex thus it must be simplified. This appendix is mainly about communications that have a wave form, in particular an EM waveform. These are possible to investigate due to existing electronic tools.

The intent of presenting the methods, concepts and mathematics used in EM signal processing is to stimulate the imagination or the researcher to use these as analogies for other channels through which communications occur. For example, the properties of the underlying reality may be directly transmitted into the world through changes in species, changes in molecules, changes in human abilities to perceive, to invent new ideas, or to produce animals with more functions.

If the researcher allows himself to experiment with such communications, he may find they are transmitted through the channel of psychic waveforms received by minds, human or non-human.

There may be a communication that yields less function due to noise in the signal. If the communication results in a newborn animal with a mutant aspect, the mutant aspect may be transient noise in the communication. Such noise temporarily produces a few mutants with, for example, a poor survival rate or no ability to reproduce. There are many channels through which the communications may take place. By proposing hypotheses and testing them, the information being communicated will be decoded.

The limited concepts of noise and these mental tools in this Appendix B can be employed to discover other messages or causes or influences from the underlying reality. The investigators can use these tools to remove the noise from the whole message, isolate the core of the message, decode the message, and make it useful to other people.

In Appendix C, several noises are described which are not wave forms; they are mental constructions, lies, and propaganda, etc. The intention is to stimulate the researcher to look for analogies not within electromagnetic phenomena and waves

Section 1. A problem with communication is to accurately determine the bare information transmitted in the messages by identifying the noise and subtracting it from the total message

Signal processing provides the basic analysis, modeling and synthesis tools for a diverse area of technological fields of work, including telecommunication, artificial intelligence, biological computation and system identification. Signal processing is concerned with the modeling, detection, identification and utilization of patterns and structures in a signal process. Applications of signal processing methods include audio hi-fi, digital TV, radio, cellular mobile phones, voice recognition, vision, radar, sonar, geophysical exploration, medical electronics, bio-signal processing and in general any system that is concerned with the communication or processing and retrieval of information. Signal processing theory plays a central role in the development of digital telecommunication and automation systems, and in the efficient transmission, reception and decoding of information.

This exposition begins with a definition of signals and a brief introduction to various signal processing methodologies. One considers several key applications of digital signal processing in adaptive noise reduction, channel equalization, pattern classification, pattern recognition, audio signal coding, signal detection, spatial processing for directional reception of signals, Dolby noise reduction and radar. This is just a beginning. It is clear that there is a wide range of electromagnetic phenomena communicated by waves. The number increases exponentially as the number of gadgets in the hands of the population increase.

This simple review of a few of the communication processes that are now common place, yields a clue about the necessity making as few assumptions as possible about the mechanism of communication from the underlying reality. Some assumptions are necessary or no investigation is possible.

In the limited range of the commercially available communication processes, the reception of information depends in an arbitrary way on a set of messages and noises with a known and combined distribution. One must ascertain how much information is delivered to the observer concerning the message alone. In any case, some assumptions must be formulated. Appendix A discusses this topic from a broader view.

A recommended project is to evaluate the methods of decoding underlying reality as far as their efficiency in transmitting information. Another approach is to decode only one type of message, for example, only search for wave representations in the information. This approach begins with the premise that most communications from the underlying reality are encoded into wave forms. The advantage is that, for processing waves, there are massive information and mathematical methods already developed and tested. This can be used as a foundation upon which to build up the description of the underlying reality.

Searching the literature on the many sorts of information decoding and on the mathematics of decoding noisy communications will accelerate the discovery of the properties of the underlying reality. Solutions of the problems of radio communications and other wave based channels of communication can indicate the proper questions to ask. The questions are an excellent point of departure for focusing on particular approaches to the problems.

Section 2. Noise and distortion defined

Noise is defined as an unwanted signal that interferes with the communication or measurement of another signal. Noise is a signal that conveys information regarding the source of the noise. The noise may interfere with the signal, or the measurement, or the perception, or the processing of the signal.

The noise may be interesting because it conveys information about the source before it is encoded and transmitted. Some noise may be the distortions introduced by the channel carrying the communication.

The noise signals integrated with any investigation of the underlying reality have many sources,

a) The human mind,
b) The physical equipment,
c) The lack of understand of how to receive the transmission,
d) The ignorance of what part of a communication is the manifestation of the underlying reality,
e) The difficulty or impossibility of separating the many complex components of the communication,
f) The human traditions of religions,
g) The extreme emotional reactions to disparate opinions,
h) The mental disease of the humans investigating,
j) The mental impairment of the humans opposing the investigation,
k) The greedy and self centered desire for fame,
l) The perversion of research to obtain money or honor,
m) The personality disorders of the people in the organization, etc.

Some of these interfere with the processing of the communication.

Section 3. Noise can be classified as follows.

1) Acoustic noise emanates from movement, vibrating or colliding things, speech, and weather.
2) Thermal noise emanates from electrons in an electric conductor.
3) Shot noise is generated in all electric conductors.
4) Electromagnetic noise is part of all electromagnetic wave phenomena as part of the background. All electromagnetic equipment generates noise especially radio, TV, electric generators and transmission lines.
5) Electrostatic noise comes from all electric voltage differences even without a current.
6) Channel distortions, echo and fading exist in imperfect communication channels such as wires, wave guides, and mobile phone transmissions.
7) Processing noise results from the equipment used to send and receive signals and to remove noise.

Depending on the frequency spectrum, or time characteristics, a noise process can be further classified into the following.

8) White noise is random signals with a flat power spectrum which may contain all frequencies in equal intensity.
9) Band limited white noise has a flat spectrum and limited bandwidth in a part of the total signal.
10) Narrowband noise has a narrow range of frequencies.
11) Colored noise is non-white or is a wideband noise with a non-flat shape. Arcane names are pink noise, or brown noise.
12) Impulsive noise consists of short duration pulses of random amplitude and random duration.
13) Transient noise pulses are long duration noise pulses.

Section 4. Factors considered when removing noise from the communication

Signal processing engineering was accelerated due to wars. Part of the advance has been in decoding encrypted secret information. There are currently many branches of signal processing. Part of the processing is reducing the problem such as was explained above, especially noise reduction. These concepts, electronic tools, and mathematical methods can be applied to the many difficult tasks in the research of the underlying reality. The most difficult task may be discovering a communication that is clearly originating from the underlying reality. The major tasks are listed below.

The history of science reveals that when one underlying cause, c_1, is determined, there is another cause, c_2, which causes c_1. An example of c_1 is that a chemical combination of two materials has different properties than either of the materials alone. The underlying concept c_2, of individual molecules combining explained this. Then the cause of the existence of different molecules was sought. One could ask what is c_3? When does the chain of cause, effect, and probability stop?

When describing diffusion, is diffusion the property of the underlying reality that is transmitted into the world? Or is there an underlying reality that influences the world to diffuse? When does one know that the manifestation of a particular reality, like sub atomic particles, is the same as the underlying reality? How does one know if the particles are the 'encoded' principle of the underlying reality or the principle of the underlying reality itself?

In science, the general procedure is to explain a phenomenon such as an animal in terms of its component organs. That is considered decoding the manifestations of the underlying reality. Another researcher examines the organs and finds cells which are considered the next level of decoded underlying reality. This indicates there are levels of underlying reality. Is the property of having several levels the manifestation of a principle of underlying reality? Then the underlying reality itself may have levels which it expresses into the world. Are there many different underlying realities that are a system manifesting into the world? Yes, there are levels of underlying reality.

Section 5. How much information does one have concerning the independent variables in a communication?

This is a basic starting point when analyzing a communication.

Wiener has presented the math description of this kind of fixed variable. Wiener gives the math for the generalized message which is embedded into the set X1, X2,…, X(n-m). The subscript, m, is the number of noises. Wiener labels the generalized noise as the set X(n-m+1), X(n-m+2),…, Xm.

The Y set is the generalized corrupted message. So one may increase the known information obtained by specifying the Y corrupted messages.

Thus if one can define

- ✓ all the noise, and
- ✓ all the assumptions of the attributes and
- ✓ all the mistaken messages introduced into the meaning of the information and
- ✓ all the dogma about the meaning of the information and
- ✓ all the corrupted messages

Then the math given by Wiener could be used to quantify the amount of information from the underlying reality. (Wiener, 1962)

Section 6. Assumptions must be framed to aid in the discovery of a reduced set of transmissions that are tractable

This problem was addressed in Appendix A. A method must be defined for the signal processing and signal decoding. There are many methods. Only bare mentions of math tools are listed next. The reader must study separate texts to decode the meaning of the many maths and how to use them.

There are many signal processing methods based on transforms. The purpose is to describe a signal in terms of a combination or set of elementary signals such as sine waves. The Fourier Transform has been widely applied to this problem of description. The Laplace transform and the z transform are similar to the Fourier transform. The wavelet transform is multi-resolution when the signal is a combination of waves of different durations. They are all used to describe wave phenomena

A model based signal processing may be invented for a specific application. It uses more information and assumptions to create a model of the signal process.

The Bayesian inference theory provides a general framework for statistical processing of random signals. Fluctuations of a purely random signal or the distribution of a class of random signals cannot be modeled by a predictive equation. They can be described in terms of statistical average values. A probability distribution function in multidimensional signal space can be used. The theory can be the basis of formulating and solving estimation and decision making problems.

The methods developed for neural networks may be useful. Neural networks are combinations of relatively simple nonlinear adaptive processing units. Arranged to have a structural resemblance to the transmission and processing of signals in biological neurons. In a neural network, several layers of parallel processing elements are interconnected by a hierarchically structured connection network. Neural networks are particularly useful in nonlinear partitioning of a signal space, in feature extraction, pattern recognition, and in decision making systems. This type of decoding is would be appropriate for living systems.

Screening all probable communications expressed in a wave function from the underlying reality is the beginning of the process. This is necessary to define a reduced set for analysis.

Defining hardware for the reception and analysis would be valuable if at all possible. There is a large variety of hardware and software in existence. This may be extremely expensive.

Section 7. Several recommended mathematical approaches

for decoding the reduced set of communications from the underlying reality

One approach to decoding the communication from underlying reality is to reduce the problem to one that obeys Figure 1 Generalized Communication and Signal Processing System.

The concept of removing noise from the communication was described for a certain reduced set of problems. Wiener gave the mathematical and logical basis for the description below. (Weiner, 1961 pp. 60-94) Below, his method has been translated into removing noise from ancient Biblical decoding of spiritual events.

There are thousands of books and technical papers about electronic communications published during the last 100 years. One topic is removing noise from a message. Two of these books are *Extraction of Signals from Noise*, (Vaĭnshteĭn, 1962) and. *Advanced Digital Signal Processing and Noise Reduction.* (Vaseghi, 2006) This Appendix addresses the subjects of these books.

There are endless books, papers and electronic gear to aid in working with the waves of the electromagnetic spectrum. A book to begin one's understanding is, *Introduction to Filter Theory.* (Johnson,. 1976) A deeper insight would be gained by designing a device to filter out the noise using *Electronic filter design handbook.* (Williams and Taylor, 1995)

Most of the interesting problems are non-linear. Therefore, *Fundamentals of nonlinear digital filtering* gives methods of decoding the non-linear subset of the communication. (Astola, 1997) An up to date literature search is necessary to keep up with the science, mathematics and engineering of wave processing. Then one can begin to process the subset of communications from the underlying reality that are embedded in waves.

Section 8. Applicable conceptual and mathematical tools for processing communications from the underlying reality and removing noise

The focus in this Appendix will be on cancelling the noise from a communication. Only the names of the math and the methods are mentioned. The reader must read the books and study the methods before they will be come meaningful.

The total observed communication, y(t), may be in the form

$$y(t) = x(t) + n(t) \qquad \text{equation (4)}$$

Where the desired clean message is x(t) and the noise is n(t). The time, t may be digital, discrete quanta, analog, t4, or other time such as the long view, t5.

There are many commercially available noise cancelling devices used, for example, in earphones. This suggests seeking a communication from the underlying reality embodied in sound waves.

The channel through which the message is transmitted may cause a distortion. The process of channel equalization is the recovery of the signal distorted by the channel, for example, with a non-flat magnitude or a nonlinear phase response. The channel response may not be known but the message can be recovered anyway.

Section 9. The signal can be classified and the pattern can be recognized

If there is a particular wave form that is expected, it can be classified as true of false by an electronic classifier which can recognize the wave form or falsify it. If there are many wave forms an electronic classifier can recognize them or falsify them in the observed signal by referring to examples embedded in the equipment.

There are many speech recognition computers. This example is not likely to be an indicator of communication from underlying reality. However, in several religious God, the underlying reality, speaks to the writer who recognizes the voice of God. Another case is a person who "channels" a God, spirit or alien. The objective in the design of voice recognition is high fidelity with as few memory bits per sample as possible to minimize memory.

Section 10. Discovering whether there is a communication in a signal that appears to be all noise

The objective is to determine whether there is non-zero information quantity in a signal or there is only noise. This is stated mathematically,

$$Y(t)=b(t)x(t)+n(t) \qquad \text{equation (5)}$$

Where x(t) is the noise free communication. If x(t) has a known wave form, a correlator can detect the communication.

And n(t) is the noise.

And b(t) is binary valued state indicator sequence such that b(t) =1 means there is a communication and b(t)=0 means there is no communication.

The impulse response from the correlator, h(t) for detection of a communication x(t) is the time reversed

Version h(t)=x(N – 1- t) where $0 \leq t \leq N-1$ and N is the duration of x(t).

This form appears to be useful because, in searching for communication from the underlying reality, most signals are mixed with many other signals. Considering the many assumptions and the noise injected into the signal by human false thinking and false beliefs, one will find it difficult to define all the relevant wave forms for the correlator to match with the multiple components of the communication.

Section 11. Example when the noise is mistaken to be the information in the communication

In ancient China, there were men whose vocation was writing things to be read and contemplating written communications. Absorption in this vocation led Liu Xie (462-522 AD) to exclude most of the world from his life except for patterns in written materials. He believed that patterns were the underlying reality. He wrote *The Literary Mind and the Carving of the Dragons* (wen-hsin tiao-lung). The contents were 38 types of literature as vehicles of truth (wen).

The word for "pattern" could have many meanings just as it does in English.

a) It could mean the natural markings of a bird or the design on a war shield and so on.
b) It could mean device or written symbol or written material or literature as an art form or the explanation of the underlying reality.
c) It could mean embellishment, refinement, culture, civilization or an example of information communicated by the underlying reality.

Liu described the patterns of stars, of flowers, of sounds and music. He believed the channel of communication with the underlying reality was through the channel of patterns. This belief may be tested and verified or falsified. (Liu, 1959)

Patterns may possibly be the noise in the communication channel through which the underlying reality transmits. Or the patterns may be the main signal when the noise is removed.

Hypotheses are proposed above to look for the communication in electromagnetic waves. This preoccupation with EM waves may possibly be noise in the communication channel. Or the EM waves may be the main signal remaining after the noise is removed.

Section 12. Amplifying only the original signal to make it more distinct from the noise

A means to discriminating between information and noise is to drown out the noise by increasing the power of the main signal. In the example when the pattern is perceived as the main signal, other parts may be the focus of attention for acquiring information. If the researcher thinks the EM waves are noise, he could ignore the EM waves and examine the dots and dashes embodied in the waves. When the EM signal of interest from the underlying reality is known to be in a certain frequency range, that range can be amplified to make it easy to discriminate from the noise.

Section 13. Restricting the original direction of the communication to limit the amount of signal

It may be useful to limit the amount of signal to a specific direction. The electronic tools are available from the specialties of radar, microwave, and petroleum exploration processing. This removes almost all noise coming from the wrong directions.

Section 14. Noise in the communication channels from the underlying reality analogous to electronic communications

The many types of noise above are characteristic for electronic communications. The various EM noises can stimulate the researcher to look for analogies to these noises outside EM sources. The mentioned noises can be extrapolated to apply to any communication from the underlying reality. The researcher can use his imagination to invent analogies to the concepts presented herein that suit investigations of the underlying reality communicated as natural laws. One analogy is the amplitude modulation of weather patterns.

Section 15. Hidden Markov Models have been used to decode communications

Markov math requires some study and concentration before it can be applied to a search for underlying reality.

Hidden Markov models are used for the statistical modeling of non-stationary signal processes such as speech signals, image sequences and time-varying noise. The Markov process, developed by Andrei Markov, is a process whose state or value at any time, t, depends on its previous state and values at time t–1, and is independent of the history of the process before t–1.

Hidden Markov Models can be programmed into the computer to allow the computer to learn: called heuristic or recursive training. This heuristic process may be useful for an oblique attempt to decode communications with the underlying reality when a direct attempt has failed. Decoding of signals using Hidden Markov Models has been used for DNA and Protein Sequence Modeling. DNA seems to be a type of communication from underlying reality. The application to DNA may be a direct interception of the communication from the underling reality to the life form.

Hidden Markov models are used for the statistical modeling of non-stationary signal processes such as speech signals, image sequences and time-varying noise. This may relate to some communications from the underlying reality that vary in many ways but are not steady.

Section 16. Decoding of signals using Hidden Markov Models (HMM)

Hidden Markov models are used in applications such as speech recognition, image recognition, signal restoration, and for the decoding of the underlying states of a signal. For example, in speech recognition, HMM are trained to model the statistical variations of the acoustic realizations of the words in a vocabulary. So one HMM for each letter sound and each word sound. Perhaps, a million HMM. In the word recognition phase, an utterance is classified and labeled with the most likely of the million candidate HMM. There would also be an HMM for silence.

Consider the decoding of an unlabelled sequence of T signal vectors, $X = x(0), X(1), \ldots, X(T-1), \ldots, X1000000$.

Where each vector, T, has many components.

There is a given set of V candidate HMM $[M1, \ldots, MV]$ which are probable matching sounds. The probability score for the observation vector sequence, X, and the model, k, can be calculated as the likelihood: $V(p)$. (Vaseghi, 2006, pp. 319ff)

Section 17. Hidden Markov Models (HMMs) in DNA and protein sequence modeling

This modeling of DNA and other chemical sequences requires fast identification of millions of probable combinations. this is impossible for a human or even a large group of humans. Computer modeling using HMM is an efficient tool.

A major application of hidden Markov models is in bio-signal processing and computational molecular biology.

a) in applications including multiple alignment and functional classification of proteins,

b) prediction of protein folding,
c) recognition of genes in bacterial and human genomes,
d) analysis and prediction of DNA functional sites and
e) identification of nucleosomal DNA periodical patterns.

Hidden Markov models are powerful probabilistic models for detecting homology among evolutionarily related sequences. Homology is concerned with likeness in structures between parts of different organisms due to evolutionary differentiation from the same or a corresponding part of a remote ancestor. HMMs are statistical models that consider all possible combinations of matches, mismatches and gaps to generate an alignment of a set of sequences.

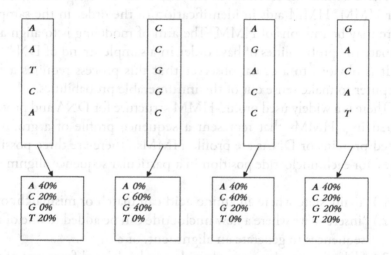

Figure B.2 A Markov model for a dataset of DNA sequences.

Each of the lower squares contains a statistical model, amount of amino acid, or Hidden Markov Model(HMM) of a known amino acid. Comparing the HMM with the DNA sample aids in identification of the amino acids in the sample. There may be millions of samples and thousands of HMM.

Figure B. 2. A Markov model for a dataset of DNA strands

FigureB. 2 shows an overly simple example of DNA sequencing; a computer based statistical modeling of DNA observations. DNA sequencing is the process of determining the precise order of nucloetides in a strand of DNA molecule. Each strand is compared with a HMM until a match is found. Each HMM identifies the strand by a probability evaluation.

The HMM method is used to determine the order of the four bases—adenine, guanine, cytosine, and thymine--in a strand of DNA, abbreviated A, G, C, T. Each of the top rectangles represents a strand of DNA. There may be millions of these. Each of the bottom rectangles represents a math model stated in percent of an order of bases A, G, C, T or HMM. HMM aids in identification of the order in the sample. There may be millions of HMM. The aim of modeling is to align and estimate the probabilities of base order in a sample strand of DNA.

It is obvious to a casual observer that this process requires a fast computer to make sense out of the innumerable probabilities.

There is a widely used profile-HMM structure for DNA and protein sequencing. HMMs that represent a sequence profile of a group of related proteins or DNAs are profile HMMs. There are three possible 'states' for each nucloetide position in a particular sequence alignment:

1.) 'main' state where an amino acid can match or mismatch, or a
2.) 'insert' state where a new nucloetide can be added to one of the sequences to generate an alignment, or a
3.) 'delete' state where a nucloetide can be deleted from one of the sequences to generate the alignment.

Probabilities are assigned to each of these states based on the number of each of these events encountered in the sequence alignment. The model allows for a transition from one state to another and is also associated with a transition probability. The greater the number and diversity of sequences included in the training alignment, the better the model will be at identifying related sequences. An adequately heuristically trained profile HMM has many uses. It can align a group of related sequences, search databases for distantly related sequences, and identify subfamily-specific signatures within large protein or DNA super-families.

Section 18. Probability mathematics is used to predict the transitions between states of a random variable

Information theory allows prediction and estimation of the history of dependencies such as the weather or an unknown message composed of symbols. In this sense, information is defined as knowledge of the states of a random variable such as the content of an encoded espionage letter.

Figure B.3 shows the power of the probability math in the simplest case, the binary condition where there are only two possible states of existence. although the notion of 'probability' suggests a lack of precision, the figure shows great accuracy when it comes to quantizing the probability itself.

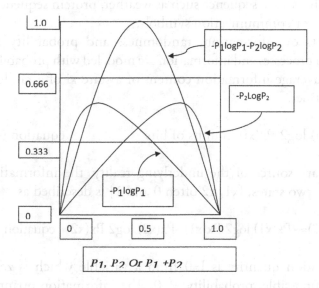

Figure B.3 Representation of Information I(x) contained in message states

P_1, P_2 and P_1+P_2 from a binary source when the maximum information content is one bit and the probability is 0.5 for each state

Probability models are the foundation of information theory used in electronic communication systems such as radar, speech recognition and noise reduction. The study and application of probability theory enables all the phone communications. Clearly the tools of probability math and information theory apply to communications with the underlying reality.

Probability models provide a complete mathematical description of the distribution of a random process. Probability models enable the estimation of the likely values of a process from noisy or incomplete observations which may be discrete or continuous information. This is appropriate for the analysis of information suspected as being transmitted from the underlying reality.

Information is knowledge regarding the states of a random variable. Information is measured in bits. One bit is equivalent to two equal-probability states, 0 or 1. Information conveyed by a random process is associated with its state sequence such as weather, protein sequences, DNA sequences, or communication symbols.

The concepts of information, randomness and probability are related. Random processes and information are modeled with probability functions. The average information content of a state xi of a random variable is quantified

As $I(xi) = -Px(xi) \log2 Px(xi)$ in units of bits   equation (6)

From a binary source of the underlying reality, the information conveyed by the two states, [x1, x2 often 0 and 1] is described as

$H(x) = I(x1) + I(x2) = -Px(x1) \log2 Px(x1) - Px(x2) \log2 Px(x2$ equation (7)

The information quantity is I=0 from a variable which is zero, meaning it is impossible, probability, P=0. The information quantity is also I=0 from a variable which always occurs P=1. Since the value is known not new information.

Thermodynamic entropy increases in proportion to the disorganization of a physical substance. Pure gold has zero entropy. As it is diluted the entropy increases. Information entropy gives a measure of the quantity of the information content of a random variable in terms of the minimum number of bits per symbol required to encode the variable.

The concept of entropy enables the mathematical description of many concepts in information theory. Entropy can be used to calculate the theoretical minimum capacity or bandwidth required for the storage or transmission of an information source such as text, image, or underlying reality. Shannon derived the entropy measure, H, as a function that satisfies the following conditions.

Entropy, H, should be a continuous function of Pi.
If Pi = 1/M, then H should be monotonically increasing function of M.

If the communication symbols are broken into two or more sets, the entropy of the original set should be equal to the probability weighted sum of the entropy of the subsets.

Section 19. Stationary and non-stationary random processes

More can be added to the subject of stationary and non-stationary random processes. Although the amplitude of a signal fluctuates with time, t, the parameters of the process that generates the signal may be stationary, time invariant, or non-stationary, time varying. An example of a non-stationary process is a series of earthquakes or typhoons in a fixed region, such as near Florida, which have varying intensity, number of waves, and spectrum of vibrations over 1000 years.

A stationary process must have time invariance of all its parameters of the probability model: the mean, the variance, the power spectral composition, and the higher order moments of the process. These parameters are the statistics, the expected values of a random process.

As introduced above, investigation of the statistical time series as a feature in the communication from the underlying reality. This may bear the fruit of discovering a communication from the underlying reality.

One could research the truth or false of the following hypothesis.

Hypothesis: There are stationary time series that continuously manifest as a physical part or a mental part of the world.

An ergodic process may be an interesting property of a signal because this may be a dominant process manifested by the underlying reality. A stationary process is said to be ergodic if it exhibits the same statistical characteristics along the time dimension of a single realization as across the space or ensemble of different realizations of the process. Over a very long time, a single realization of an ergodic process takes on all the values, the characteristics and the configurations exhibited across the entire space of the process.

In general, this is stated mathematically,

statistical averages [x(t, s)] across time and space =statistical averages [x(s,t)] across space and time   equation (8)

Hypothesis: There is a class of communications from the underlying reality that is ergodic.

Section 20. There are many mathematical and electronic tools available for investigating the communications which take the form of waves

Clearly, there are many mathematical and electronic tools that can be applied to the search for the many types of communications from the underlying reality that take the form of waves. There are many more tools such as Bayesian inference, interpolation, least squares model fitting and adaptive filters.

Adaptive filters are applied to three broad signal processing problems as follows.

a.) Removing noise and channel distortions.
b.) Tracking and prediction of a non-stationary signal observed in noise.
c.) System identification by an estimation of the parameters of a time varying system from an observation.

Adaptive filters work on the principle of minimizing the mean squared difference (error) between the filter output and a desired signal. Adaptive filters are used for estimation of non-stationary signals and systems. Applications of adaptive filters include multichannel noise reduction which can be assumed to be a property of communication from the underlying reality. The Kalman filter theory has a state equation that models the dynamics of the signal generation process. It has an observation equation to model the channel distortion and the additive noise. These are properties of communications from the underlying reality. By testing various hypothetical models against data and removing channel distortion and additive noise, one can experiment with many models until finding one that reproduces the way the underlying reality generates the communications that are manifest in the world.

Another tool is spectral amplitude estimation when researching aspects of nature with amplitude modulation such as typhoons, earthquakes, the rise and fall of civilizations, epidemics, and the archetypal expressions of humans such as war, hierarchy, and cycles of destruction of the earth's surface.

Section 21. Is impulsive noise a valuable topic of research?

The human nervous system contains neurons, a channel for communicating information. Part of the channel is electro-chemical pulses in the form of waves which migrate to dendrite of the individual neuron. The pulse may be discharged across the synapse into the next neuron depending on the summation of the inhibiting and transmitting pulses. This suggests that this is a location to observe the communications, decode them, and investigate the information content. (Popper & Eccles, 1977, p. 227)

Impulsive noise within the neural waves consists of relatively short duration on/off pulses caused by many sources such as the channel environment or the addition of pulses from other neurons. An impulsive noise filter can be used for enhancing the quality and intelligibility of noisy signals and for achieving pattern recognition. Optimal performance is obtained by using the following.

a) Distinct features of the noise of the signal in time or frequency domains
b) Statistics of the signal and the noise processes
c) A model of the constitution of the signal and noise generation.

Section 22. Is transient noise a promising approach to decoding the communications from the underlying reality?

Transient noise pulses differ from the short-duration impulsive noise studied in the previous section, in that they have a longer duration and a relatively higher proportion of low frequency energy content, and usually occur less frequently than impulsive noise. Next, one considers the modeling, detection and removal of transient noise pulses.

The sources of transient noise pulses are varied, and may be electromagnetic, acoustic or due to physical defects. An important observation in the modeling of transient noise is that the noise can be regarded as the impulse response of a communication channel, and hence may be modeled by one of a number of statistical methods used in the modeling of communication channels.

The initial pulse is due to some eternal or internal impulsive interference whereas the oscillations are often due to the resonance of he communication channel excited by the initial pulse. It may be the response of the channel to the initial pulse. A noise pulse originates at some point in time and space and then propagates through the channel to the receiver. Transient noise pulses often consist of a relatively short sharp initial pulse followed by decaying low-frequency oscillations.

Section 23. Transient noise pulse models

To a first approximation, a transient noise pulse, $n(m)$, can be modeled as the impulse response of a linear time invariant filter model of the channel as

$$n(m) = \sum_k h_k \sum_j A_j \, \delta[(m-T_j)-k] = \sum_j A_j \, h_{m-T_j} \qquad \text{equation(9)}$$

where it is assumed that the jth transient pulse is due to an impulse of amplitude A_j at time T_j.

A noise model should be able to deal with variations. There are three methods of modeling the temporal, spectral and durational characteristics of transient noise as follows.

1) A template based model

2) A linear-predictive model
3) A hidden Markov model

Section 24. Removal of noise pulse distortion

One considers two methods for the removal of transient noise pulses:

(a) an adaptive noise subtraction method; and
(b) an autoregressive model-based restoration method. The noise removal methods assume that a detector signals the presence or the absence of a noise pulse, and provides additional information on the timing and the underlying states of the noise pulse.

Section 25. Adaptive subtraction of noise pulses

Let $x(t)$, $n(t)$, and $y(t)$ denote the signal, the noise pulse and the noisy signal, respectively; the noisy signal model

is $y(t)=x(t)+b(t) n(t)$ equation (10)

Where the binary indicator sequence, $b(t)$, indicates the presence or the absence of a noise pulse. Assume that each noise pulse, $n(t)$, can be modeled as the amplitude-scaled and time-shifted version of the noise pulse template

ň(t), so that $n(t) \approx w \, ň \, n(t-D)$ equation (11)

where w is an amplitude scalar and the integer D denotes the relative delay (time shift) between the noise pulse template and the detected noise. From equations (10) and (11) the noisy signal can be

modeled: $y(t) \approx x(t)+w \, ň \, n(t-D)$ equation (12)

From Equation (12) an estimate of the signal, $x(t)$, can be obtained by subtracting an estimate of the noise pulse from that of the noisy signal:

$x(t)=y(t)-w \, ň \, n(t-D)$ equation (13)

where the time delay D required for time-alignment of the noisy signal, y(t), and the noise template, ň(t), is obtained from the cross-correlation function.

Section 26. Channel equalization and blind de-convolution

Blind de-convolution is the process of unraveling two unknown signals that have been convolved. An important application of blind de-convolution is in blind equalization for restoration of a signal distorted in transmission through a communication channel. Blind equalization is feasible if some useful statistics of the channel input and also of the channel itself, are available. The success of a blind equalization method depends on how much is known about the statistics of the channel input.

Section 27 Noise in wireless communications

This suggests pondering the analogy to the cell phone transmitter-receiver towers when experimenting with communications from the underlying reality. The various problems solved in wireless communications can be applied.

Hypothesis: Each part of an animal or inorganic entity has many cells that receive slightly different information from the underlying reality.

There may be an analogy between the cell phone transmission and reception network and the communications from the underlying reality to the many humans or other large numbers of animals and plants. Finite radio frequency channels among many different mobile users and for many different applications and purposes. The available power is limited by the capacity, size and weight of the on-board batteries. In a mobile communications system the radio frequency bandwidth needs to be used efficiently in order to maximize the capacity.

Imagine the geometric topology of a cellular mobile radio system. In cellular systems a city or a town is divided into a number of geographical cells. The cells are thought of as having a hexagonal shape. A key aspect of the cellular mobile communications technology is the very large increase in spectral efficiency achieved through the arrangements of cells in clusters and the reuse of the radio frequency channels in non-adjacent cells; this is possible because the cell phones and base stations operate on low-power transmitters whose electromagnetic wave energy fades before it reaches non-adjacent cells. In the cellular configuration, each cell within each cluster of seven cells uses a set of different frequencies (each set of radio channel frequencies used in a cell is a different number). Different clusters reuse the same sets of frequencies. Each cell has a base station that accommodates the transmitter and receiver antennas, the switching networks and the routing equipment of the base station.

Consider the analogy,

a) there are the base transmitter-receiver antennas,
b) the base station controller is responsible for allocation of the radio channels, channel management, encoding of the channels onto wired channels and execution of the hand-over function as the transmissions arrive from different directions,
c) there is the operation and maintenance center, a database holding information about the overall operation and maintenance of the network of communications.

The similarity between

1) processing functions in a communication system and
2) hypothetical processing functions of the communication from the underlying reality

includes the following elements.

d) Source coder and decoder which compress signals at the transmitter by removing excess parts of the message. This can use signal transforms such as discrete Fourier Transforms or signal probability models such as entropy models.
e) Channel coder and decoder which reduce transmission errors due to noise, fading and loss of parts of the information.
f) Multiple access signaling provides simultaneous access to many communications on the same shared resource.
g) Cell handover determines the direction from which the communications are received.

h) Channel equalization removes the distortions and time dispersion of signals that result from the characteristics of channels.

j) The frequency range is used for all cells. This increases the capacity of he channels. The cells have weak signal power and thus the same frequency does not overlap between cells.

Other multipath distortions and noise may occur. The wave may be reflected from a smooth surface. The wave may be diffracted by a large dense object. The wave may be scattered by an uneven surface with large dimensions or by an object comparable in dimension to the wavelength of the signal. Thus, the transmitted signal may arrive at the receiver from different directions and at different times. This can reduce the capacity of the channel or even destroy the signal.

Section 28. Analyzing the power spectrum to find signal and remove noise

The energy or power spectrum analysis is concerned with the distribution of the signal energy or the power in the frequency domain.

Consider a typhoon as a communication from the underlying reality. It has enormous measurable energy. It is composed of many frequencies, each of which has measurable power. There are endless examples of natural phenomena with energy and power. One could analyze the power spectrum to find the signal and remove noise.

If a phenomenon can be defined with a deterministic time signal, the energy spectral density is well defined mathematically. The math tool is the Fourier transform of the autocorrelation function of $x(t)$.

Unfortunately, the Fourier transform exists only for a finite energy signal. Therefore the class of signals from the underlying reality which can be expected is a stationary stochastic signal. Since these are stationary, they are infinitely long and have infinite energy. Therefore they do not have a Fourier transform. For these stochastic signals, the quantity of interest is the power spectral density, defined as the Fourier transform of the autocorrelation function of the power spectral density.

The power spectrum reveals the existence or absence of repetitive patterns and correlation structures in a communication process. There is a common electronic tool, based on the Fast Fourier Transform, that yields various desired outputs such as the Fourier Transform and the correlation structures. Other tools yield better frequency resolution and less variance. The correlation is a function of the time domain. It reveals the information on periodic or random structure of a communication.

The power spectrum is a function of the frequency domain. It gives the distribution of the power among various frequencies and shows the existence and relative power of repetitive patterns and random structures in a communication.

There is an uncertainty principle because of the error in frequency determination

$$\Delta f = 1/\Delta t \qquad \qquad \text{equation (14)}$$

Where Δt is the error in the time determination. So decreasing the frequency error increases the time error.

Section 29. Model based power spectrum estimation

One could analyze the signal in a case when one knows the parameters within the signal. But this is not usually the case. Thus models are invented. With a model one can extrapolate beyond the range where data is available. If the model is translated into linear mathematics, the signal, x(t), can be considered the output from a linear time invariant underlying reality excited with a random flat spectrum excitation. This assumes that the power spectrum of the output is shaped entirely by the frequency response of the model. The input-output relation has been deduced.

Section 30. Maximum entropy spectral estimation

The quantity of information in a communication is the negative of the entropy in math terms. The maximum entropy principle is appealing because it assumes no more structure in the correlation sequence than that indicated in the measured values. The math of the maximum entropy power spectrum estimate using models has been deduced. Other deductions can be made when one applies the theory of communication to the problem of decoding the measured values of the communication. By investigating the maximum entropy in combination with the theory of communication, one can glean more information useful for feedback to invent a better model.

Section 31. Clues to decoding communications from the underlying reality suggested by advanced digital signal processing and noise reduction

The assumptions and limitations in Appendix A reduce the challenge of describing the underlying reality to searching for communications in wave form. This simplification reduces the problems substantially because wave forms have been studied for hundreds of years. This has produced a vast set of conceptual, mathematical, and electronic tools which can be brought to bear on the challenge.

Section 32. High resolution spectral estimation based on subspace eigen-analysis

The math of eigen-analysis is well developed. Eigen value analysis has proved effective in many types of analysis. Eigen-analysis can be used to estimate the values of parameters of sinusoidal signals in additive white noise. One of the many uses for this tool, eigen-analysis, is to use it to partition the eigen-vectors and the eigen-values of the autocorrelation matrix of a noisy signal into two subspaces as follows.

1) The signal subspace composed of the principle eigen-vectors associated with the largest eigen-values.
2) The noise subspace represented by the smallest eigen-values.

One could assume that any communication from the underlying reality would consist of many simultaneous signals. The multiple signal classification (called MUSIC software) algorithm is an eigen-based subspace decomposition method for estimation of the frequencies of complex sinusoids observed in additive white noise.

Section 33. The approach of limiting a search to wave based communications with amplitude modulation

The definition of information with random static noise and other types of noise is necessary. The message can be restricted to a definite frequency range and power output in this range to improve the probabilities of decoding the message.

No means of information transmission by radio is as efficient as amplitude modulation (AM). But humans use electronic tools to receive AM radio transmission. One could ask, "Does the underlying reality modulate the amplitude of a message?" This may be fruitful to seek a means of AM in the expression of the underlying reality.

When does the expression of underlying reality have a measurable variation of amplitude? How can this variation be recorded for human study? The weather, seasonal temperature variations, and food crop production are examples. This may be fruitful to answer, "Is it feasible to develop a means of AM in the expression of the underlying reality?" Many earth events have a frequency and amplitude which allow accurate decoding. It is recommended that these AM conditions be researched to improve the architectural and Feng Shui designs by including the probabilities of longer historical influences on the building, park or city.

Incidentally, the steel structures of high rise buildings have indeed been analyzed using the eigen-vector approach. If amplitude modulation is introduced theoretically into a building structure, then destructive vibration modes are revealed. This could be examined as a communication from the underlying reality.

One could ask, "Does the underlying reality modulate the amplitude of a message?" This may be fruitful to seek a means of AM in the expression of the underlying reality such as the building structure.

When does the expression of underlying reality have a measurable variation of amplitude? This would be a fruitful region to explore for decoding the underlying reality.

Section 34. Human extrasensory perception may be an effective tool in decoding the communications from the underlying reality

The key to understanding science and math is that they are inventions of the human mind. Consider the notions of time and space. They exist mainly in the human mind. There are many arguments about time and space even if one ignores that Einstein combined them into one concept. One serious debate is to model time as a continuous flow or to plot a given event within a five dimensional grid of reality when the fourth and fifth dimensions are times.

The notions of past, present and future are not the same for all physicists. Most math descriptions of reality in the physical sciences do not require that time changes into the future but allow time to regress into the past. Clearly there is not a well defined decoding of the expression of underlying reality in space-time. This is an example of how human perceptions have corrupted the message from the underlying reality about time. (Davies and Gribbin, 1992, p. 134)

Some examples of events with modulation of amplitude follow.

a) Hurricanes in the same region
b) Human behavior that is repetitive such as war, or appointing hierarchies.
c) The population of a species of plant or animal increases, decreases, and becomes extinct. The Feng Shui expert could observe the cycle of over- population and starvation due to the placement of mountains, valleys, agricultural fields, the waves in a stream, and the frequency of rains alternating with droughts.
d) Rain and correlated rise of river water level. The Dujiangyan Irrigation Project is an example for the decoding of this Amplitude Modulation.

Section 35. Dujiangyan Irrigation Project: an example of Feng Shui design lasting 2200 years

In Chinese culture, chaos is the opposite of the ideal condition, stability.

There is a specialty in mathematics called Fourier analysis that can describe some apparently chaotic phenomena. This analysis sums up the many component waves to yield the overall resulting wave. One can observe phenomena that are composed of processes that conflict with one another and later cooperate with one another. This is usually the source of perfectly repeating wave motion. This is also the source of near wave motion in chaos.

An expert in Feng Shui could recognize the inherent chaos or wave motion in a region if features were adjusted to aid the chaos or wave motion. Then adjustments could be made to prevent catastrophic chaos. This occurred in 200 BC in China.

There exists a water control project in China where the architectural engineer could detect the yearly floods and droughts and also the fluctuation of these floods and droughts over the duration of centuries. This is reported in *Controlling the Dragon: Confucian Engineers and the Yellow River in late Imperial China*. It relates the story of humans coping with the Yellow River for thousands of years. The focus is on the years 1495 to 1855. In 1855 the problems faced by China as a whole country overpowered its ability to continue to contain the Yellow River and especially its silt. (Dodgen, 2001)

In China, Chengdu is always praised as the Tian Fu Zhi Guo, which means 'Nature's Storehouse'. Over 2,200 years ago, the city was threatened by the frequent floods caused by flooding of the Minjiang River, a tributary of the Yangtze River (Mandarin: Wan Li Chang Jiang). Li Bing, a local official of Sichuan Province at that time, together with his son, decided to construct an irrigation system on the Minjiang River to prevent flooding. After a lengthy study of the fluctuation of these floods and droughts over duration of centuries, they designed the most effective river control in history. This was a kind of Feng Shui. Massive hard work by the local people achieved the great Dujiangyan Irrigation Project. Since then, the Chengdu Plain has been free of flooding and the people have been living peacefully and affluently for 2200 years. Now, the project is honored as the 'Treasure of Sichuan'.

Dujiangyan is the oldest and only surviving river control and irrigation system in the world without imposing a dam on the natural flows. This is an example of the early application of Feng Shui, an excellent development of Chinese science. The project consists of three important parts, namely, Yuzui, Feishayan and Baopingkou scientifically designed to automatically control the water flow of the rivers from the mountains to the plains throughout the year. (VanSlyke, 1988)

Section 36. Information theory and Time series mathematics is often applied to slowly changing and extremely rapidly changing sequences of measurements

The description of time series phenomena is a common goal in engineering and physics. After mathematically describing the phenomena, the noise must be removed. The general approach to a simple situation is shown next. An example of this approach is supplied in the next section. This is the precise application of the entropy law of thermodynamics in communication engineering. Conversely, the greater specification of an ambiguous situation, on average, will generally gain information and never lose it.

The amount of information in a signal is related to the concept of entropy in thermodynamics and statistical mechanics. Entropy is the measure of the amount of disorganization in a system. In communication, the amount of information is a measure of the amount of organization. One is the mathematical negative of the other.

There is a large class of phenomena in which what is observed is a numerical quantity or a sequence of numerical quantities, distributed in time. Temperature [or any measurement] as recorded by a continuous recording thermometer or weather data are all time series, continuous or discrete, simple or multiple. Their study belongs to the more conventional parts of statistical theory. Time series mathematics is often applied to slowly changing and extremely rapidly changing sequences of measurements. (Weiner, 1961. p. 60-85)

Slowly or rapidly changing sequences of voltages in an electronic device can be analyzed by statistical methods. The analysis must be as rapid as the sequence of signals. They may be frequency modulating networks. The receivers are quick acting arithmetical devices.

They are constructed with all the capabilities necessary for analysis; recording, preservation of memory, transmission, and preparation of the analysis for human or machine use.

The mathematics processing of the information can be written when the information is known exactly but perfectly known information is almost never received. An example of the math used in a situation when there is an error yields a description of the quantity of information. A simple form of information is the record of a choice when there are only two possible choices: heads or tails of a coin toss. The choice is called a 'decision.'

One could ask for the amount of information in a perfectly precise measurement of a quantity known to lie between A=0 and B=1. The measurement has an equal probability of being anywhere in the range. The measurement is represented by an infinite number, a1, a2, a3,..., an. Each binary 'a' has a value of zero or one. The amount of information is infinite. If there is a uniformly distributed error lying over a range b1, b2,..., bn, where bk is the first non-zero number, then all the decisions between ak to a(k-1) are significant. The number of decisions made is not far from

$$Q= -\log_2 (b_1 \times b_2 \times ... b_n)$$

This is the quantity of the precise formula for the amount of information and its definition.

One could determine the amount of information gained by fixing one or more variables in the problem. Let u=message and let v= noise.

Let w=u+v= constant

The information contained in a precise message with no noise is infinite. If noise is present, then the information is finite. The information is zero if noise is significant.

Information has the properties of entropy. The processes that lose information are closely analogous to processes that gain entropy. These processes are fusion of regions of probability which were originally distinct.

No operation on a message can gain information on average. This is the precise application of the entropy law of thermodynamics in communication engineering. Conversely, the greater specification of an ambiguous situation, on average, will generally gain information and never lose it.

Consider the case when there is a probability distribution with a n-fold density f(X1, X2,..., Xn) over the variables X1, X2,..., Xn. Let the probabilities of what the underlying reality is communicating be expressed by probability densities f(X1, X2,..., Xn) indicating n-fold specific, countable, and related types of meanings.

Let there be m dependent variables, Y1, Y2,..., Ym. How much information do we get by fixing the dependent variables?

Let Y1 have a tiny range of possibilities
between y1* and y1* + dy1*.
Let Y2 have a tiny range of possibilities
between y2* and y2* + dy2*.
The same probabilities can be conjectured for all Y3, Y4, and Y5.
Let the other dependent variables also be restricted to a tiny range.
The tiny range of Y4 is between y4* and y4*+dy4*.

Let Ym have a tiny range between ym* and ym* + dym*.

The amount of information can be expressed algebraically in a long formula.

How much information can one calculate about the variables X1, X2,..., Xn Alone?

This is another long formula.

Let the set X1,..., X(n-m) be the message. Let X(n-m+1),...,Xm be the noise.

Let Y1, Y2,..., Ym be the corrupted set; the sum of mian message and noise.

An algebraic formula can describe the general solution to the calculation of information. (Weiner, 1961, p. 60-66)

This is the precise application of the entropy law of thermodynamics in communication engineering. Conversely, the greater specification of an ambiguous situation, on average, will generally gain information and never lose it.

Section 37. An ancient example of decoding communications from the underlying reality and suggestions to remove the noise

How is information measured? A simple case is the choice between two equally probable alternatives when one or the other is bound to happen. Another case is defining the information gained when one or more variables in the measuring system are fixed. Another case is when there is noise in the signal.

Let n=message from the underlying reality

And let m=noise which could be the human delusional insistence on preconceived notions, or the misinterpretation of what the underlying reality is transmitting, or poor observation skills or dishonest intent of the observer and so on. If m=0 then n contains infinite information from underlying reality. When there is noise, information approaches zero rapidly as the noise increases.

Consider the case of fixing the dependent variables. This would require a better set of hypotheses about what is clearly a communication from the underlying reality.

An example of a persistent error happens when there is disagreement. A set of people to insist that events or physical laws are caused by underlying reality. Simultaneously another set of people are convinced that physical laws are the underlying reality itself. It is necessary that there are minimum assumptions of the probability of correctly decoding the observations received and so on.

Let the probabilities of what the underlying reality is communicating be expressed by probability densities $f(X1, X2,..., Xn)$ indicating n-fold specific, countable, and related types of meanings. Then the laws of probability apply.

Let the n independent variables of the underlying reality be X1, X2,..., Xn. These are the actual attributes of underlying reality.

This is a concrete example which has not been adequately doubted. The underlying reality is labeled, "the Lord."

There exists a set of noisy decodings of the attributes of "the Lord" introduced in the Christian Bible at the book of Kings which explained why Solomon lost his king position. (Anonymous, 1910a) The Jews who wrote the Bible were convinced the Lord had specific attributes.

As one reads Kings, one can notice the attributes such as

X1= the Lord's possessiveness,

X2=His jealousy,

X3=His anger,

X4=His drive for revenge,

Let X5=He is kind to people who pray to Him and so on.

These were traditional features attributed to the Jewish Lord.

Let there be m dependent variables Y1, Y2,..., Ym such as outcomes that are being prayed for. The outcomes were imagined to be decided by the Lord. This meant that the Lord communicated the outcome. Propose some outcomes as follows.

Y1=find a lost man;

Y2=spontaneously make a killer into a saint;

Y3=free a man from disease;

Y4=the Lord only punishes men who do not walk in the ways of the Lord;

Y5= the Lord's judgment is influenced by everyone's being purified and praying fervently;

Ym=the accuracy of the measuring instrument that is decoding the Lord. The instrument was a priest listening for the words of the Lord. He is the only one who can hear the words.

How much information does one get if one or more of the dependent variables, Y1 etc. are fixed or restricted to a tiny range of probabilities?

Let Y1 have a tiny range of possibilities between y1* and y1* + dy1*. This could mean that the missing man is nearby.

Let Y3 have a tiny range of possibilities between y3* and y3* + dy3*. This would mean the man is seeking a higher power to cure him.

The probabilities can be conjectured for all Y2, Y4, and Y5.

Let Ym have a tiny range between ym* and ym* + dym*. This would mean that the priest is using highly accurate extra sensory perceptions polished by practice, purification, fasting and chanting. He hears the Lord accurately.

Let the other dependent variables also be restricted to a tiny range. The crisis which is the focus of the Biblical text in Kings is presented in Y4. The tiny range of Solomon's behavior is between y4* and y4*+dy4*. All the people except Solomon have been purified by praying for three days so that there is probability approaches zero that the Lord would have punished them. But the Lord severely punished Solomon by taking his kingdom. Solomon behaved outside the tiny acceptable range of Y4

Employing the scientific method, some people would believe that all the dependent and independent variables carried too much noise of preconceived notions, arbitrarily invented conditions, a dishonest priest and a tradition based in emotion, not scientific observation. Most Christians and Jews would assume that the words in the Bible are 100% true. Thus, it is impossible for everyone to agree on a way to discover a verifiable or falsifiable interpretation of what the Lord encoded into the final punishment. What was the independent variable in the Lord that was transmitted? The Lord communicated the event of Solomon losing his kingdom.

This is a repetition of the example of a persistent error happens when there is disagreement as stated above. A set of people insist that events or physical laws are caused by underlying reality, the Lord. Simultaneously another set of people are convinced that physical laws are the underlying reality itself.

A scientific method of decoding Solomon's loss of kingdom could be constructed by leaving out all the noise: the guesses and false convictions, the preconceived and emotional attributes, the expectations of answered prayers and so on. None of these ancient ways of interpreting reality and Lord's purpose can be reproduced to yield a useful result at another time and another situation. After subtracting the noise, read the pure dependent variables and read the information transmitted from the Lord without noise.

This is an application of the thesis of this Appendix: remove the noise from the total corrupted communication.

Section 38. Another ancient view of reality from the book of Job in the Bible

"There was a man in the land of Uz whose name was Job and that man was perfect and upright and one that feared God and eschewed evil." God tested Job's faith, loyalty to his religious convictions. Satan tested God's convictions about reality. God and Job tested Satan's conviction of reality. This book stimulates difficult feelings in the reader. There are several convictions of the reality which the reader must wrestle with. What are God and Satan? Did Job pretend to accept the destructions God subjected him to or did he really believe God was always correct? Did Job ever deny that he himself was perfect, upright and that God was not punishing him? When Job admitted he did not understand the ways of God, God rewarded him. (Anonymous, 1910b)

Many books have been published which discuss the convictions and the meaning of the book of Job. Job raises deep questions about the source of reality which is the Christian and Jewish God, and their relation to the creation of man. (Acocella, 2013)

Section 39. Another problem of communication engineering is to know how much information is given by observations about the message alone

What is the final residual of information when all noise has been removed? The researcher must determine what levels of reality are being revealed. What are the limits of the domain of noise?

Finally, what are the attributes of the source of underlying reality before it is encoded and transmitted into the universe?

When receiving the signal from the underlying reality, there are several other major problems. Some relevant inquiries follow.

What is the whole communication, message, noise, corruption?

What are the channels of information?

What forms do the communications take? Language, symbols, physical laws, psychic reception within a human mind?

How does one differentiate between the communication and the world produced by the communication?

How many causal links are in the chain from underlying reality and the effects in the world?

There are three or more major types of communication

1. Amplitude modulation,
2. Frequency modulation,
3. Phase modulation.

Are these used in communications from underlying reality? What kind of decoding devices can be used? How can one transform the communication into one suitable for reception by human senses?

How can the communication from the underlying reality be decompressed?

There are many kinds of information compression in use.

There is a message type which is homogeneous in time, a time series, which is in statistical equilibrium. It is a single function or a set of functions of the time, which forms one of an ensemble of such sets with a well defined probability distribution, not altered by the change of t to t +T throughout. The time series is in statistical equilibrium when 'n' changes into n +V (V is an integer). This is equivalent to a measure preserving transformation into itself of the interval (0 to 1) over which the parameter A, in f(A), takes all its values.

In the case of an ensemble of functions f(t), except a set of cases of zero probability, one can deduct the average of any statistical parameter of the ensemble from the record of any one of the component time series by using a time average instead of a phase average. One needs to know only the past of almost any one time series of the class. Given the entire history up to the present of a time series known to belong to an ensemble in statistical equilibrium, one can compute with probable error zero the entire set of statistical parameters of an ensemble in statistical equilibrium to which that time series belongs. Based on logical concepts, one can compute the whole amount of information which knowledge of the past will give us of the whole future beyond a certain point. This logical fact can be applied usefully in discovering the past.

Investigation of the statistical time series as a feature in the communication from the underlying reality may be fruitful. There may be a stationary time series that continuously manifests as a physical part or a mental part of the world. An example is ocean waves. Another is the rotation of the Earth and the moon.

The time series above are simple time series in which a single numerical variable depends on the time. There are also multiple time series in which a number of such variables depend simultaneously on the time. These are of the greatest importance, for example the weather map. In this case one has to develop a number of functions simultaneously in terms of the frequency and the quadratic quantities are replaced by pairs of arrays; matrices.

The statistical series above require the knowledge of the entire past history of the time series. This is not practical. However, Heisenberg and the quantum physicists based much of their conclusions that the past does not predict the future but influences the distribution of the possible futures of the system.

There is a range of precision when an approximate knowledge of the past is experimentally acceptable. The theories of entropy and chaos explain how the time series transforms into another state. The states which it transforms into depend on the number of possible states and the probability that each state will be realized. It may be that genes, viruses and proteins transform into various states. This could be researched using Hidden Markov Models pointed at in Section 17 above.

This phenomenon may a fact of existence communicated into the world by the underlying reality. Or the transformation may be a property of the underlying reality. In quantum physics, the concept of a particle is a probability wave. The particle many not be a point object but a smear of reality. Below 10-34 joule-sec, reality is not a deterministic process. This is a numerical limitation on the research.

Section 40. Conclusion about removing
noise to decode the information

After locating a site with the characteristic that enables it to be included in the set of sites communicating from the underlying reality, all the above noises, interferences and distortions may be part of the total system which must be removed, or minimized before it is possible to decode the signals to yield useful information.

Appendix C: Many Different Varieties of Reality

There are almost infinite realities: each person has several, each group has one, each profession has one. Each human mind invents a reality which represents aspects of its associated body and its accumulation of experience and mental creations. It is impossible to list all realities even those that can be found in written form.

Each of the references at the end of the book describes one or more perceptions of reality or tools used to interpret reality.

This Appendix lists diverse realities in brief and truncated descriptions. The reader must research each one in depth to become familiar with a reality which interests him.

Section 1 The Buddha taught the detailed method of perceiving reality

The Buddha established a set of assumptions of reality called 'correct view.' After accepting the correct view, the monk pursues a method that removes all influences that cause the monk to deviate from the realization of the underlying source of reality. The final realization of the underlying source is called 'liberation from the taints of the ordinary untaught mind.' It is also termed 'enlightenment.'

The Buddhist view is that the ultimate reality is realized when one has stopped elaborating on the sense data and on the ideas one has aggregated. Nirvana is the reality one seeks to attain. What is nirvana? It is the embodiment of truth. How does one enter nirvana? One enters by the total extinction of conceptions of self, extinction of what belongs to self and other practices. It is the stopping of body-mind phenomena from appearing as though they were reality itself.

The Tibetan Buddhists have elaborated on these teachings. (Dalai Lama, 2012, p. 143 - 157)

A discussion arose about emptiness as a concept upon which to base teaching methods that lead to enlightenment. One of two conflicting views of emptiness within Tibetan Buddhism is Prasangika. This states that reality is composed only of what is independent of all other things and processes. Reality originates independently. It teaches that things

and events that depend on other things and processes for their existence are not included in the set called 'reality.'

Therefore, almost everything that is perceived as existing objectively by humans and animals is not part of underlying 'reality.' There is no ground, no self supporting basis, nothing to grasp, nothing to cling to within the set of underlying reality. All phenomena are dependent on other factors such as dependent on the language, and on the origination of later language defined by earlier language definitions which are dependent on even earlier language. There are layers of definitions. A monk realizes that there is no intrinsic reality but there is actual existence based on the origination of things depending on other things. What is termed 'objective reality' is a human experience of sense data also called World1. The sense data is labeled by language about what is experienced due to earlier sense data. Language and mental streams elaborate the sense data also called World2. But there is no existence of things that are independent of other elements of the world. This realization is labeled, "emptiness of things."

Note that emptiness was the same assumption upon which modern physics was based. The physics of emptiness is discussed elsewhere in this book.

A conflicting view of emptiness within Tibetan Buddhism is Svatantrika. In this view, there is inherent objectivity in things and processes that most humans agree on and believe exist as sense data, World1. In this view, things and processes exist in the human system of human perception and language undistorted by the mind.

There is a pervasive mental habit of accepting that the self has a concrete reality and that it is independent of other causes and influences in the environment. A person is an aggregate of other elements: food, air, feelings, and tendencies. One believes that thinking, differentiating between oneself and another object, and feeling are evidence of existence apart from the environment. But all the elements and the aggregate of elements depend on other influences and events for their existence.

The foundation according to the Buddhist view is composed of whatever is not dependent on other things, influences, and processes. That which is not dependent on other things and processes is part of the underlying reality. There are hundreds of printed lessons teaching the mental preparation for intercepting the communications of the underlying reality, labeled, Buddha nature. The communications and the embodied information are called Dharma. (Buddha, 500BC)

These two opposite views of reality can be held simultaneously. In the Svatantrika view, each person creates his world from sense data. But there are only 6 basic senses if one includes psychic awareness. These senses are very limited. They exclude the majority of possible information such as what is happening 10 kilometers away. Humans can see a tiny slice of the electromagnetic spectrum that is called, 'light.' Most of the rest of the electromagnetic spectrum, x-rays, radio waves, cosmic rays, etc. can only be detected with instruments. The world of a person includes small pieces of the whole physical world. No person creates the perception of the whole world as a single organism. To the contrary, there is an arbitrary separation of the world parts, animals, plants, moving things, rocks, without end. But it is obvious that all things and processes depend on other things and processes to exist. A baby depends on two parents to be born. The three of them can be conceived as a family, one thing. A farm depends on plants, animals, equipment, humans and so on to exist. A farm could be conceived as a whole thing not in pieces. A human cannot exist without the earth. The earth can be conceived as one thing going through a process of change.

A person divides up the sense data when it is actually an integrated whole. There is the mental conception of atomic particles which are never sensed. People and animals are thought to have egos and thoughts. Death is arbitrarily pronounced to be evil; life is good. Humans artificially spread their egos over the earth thinking they own the earth and can arbitrarily break it up without killing the earth. Often, a single human believes he can leave all society, all plants, all animals, and live somehow independently in outer space. These are not scientifically proven notions. These are delusions.

A person believing the Prasangika view observes that one human will die soon after trying to be completely independent. The whole earth organism is being killed by processes which include people. The ego is a mental invention; not possible to prove its existence. All sense data must be integrated into a whole view of the world; not arbitrarily broken into pieces. There are endless processes interacting. Not all processes are the result of cause and effect; some are probable events composed of large regions of the world. Consider a nation. It cannot exist without processes of diffusion of goods and services from outside its imaginary border. Each good or service is not defined separately because it is a process composed of many things diffusing through the nation. The language is import-export. The sight from above the earth in an airplane does not reveal separate national boundaries. No individual nation can be detected. The nation depends on other things outside its arbitrary borders. It depends on language and human imagination.

It is possible to hold both views of emptiness simultaneously: Prasangika and Svatantrika. Realizing emptiness is one step in achieving liberation from the taints of the ordinary untaught mind and achieving enlightenment in detail. A given person is constantly changing; not a stable unit. A human can evolve into a liberated state and can abandon the habit of defining the world into parts.

In addition to realizing emptiness, an individual or group can live in a different mentally invented state of being such as an emotional state or a conviction about what his world consists of. A person can pursue mental entities such as described in Appendix D: An Attempt to Label Some of the Separate Levels of Mind and Consciousness. Then a person may be able to exist in a state of higher level of mind, the person goes beyond the above two views. He may exist in a state of neither perception nor non-perception.

Section 2. The quantum physicists taught the detailed method of perceiving reality as emptiness and as probable, existence of processes

The early researchers into the science of physics practiced the mental habit of accepting that the human self is identified as an independent thing in the world and that the self is real and exists by itself. This habit was reflected in accepting that sense data experienced in a laboratory represented a concrete reality. This habit remains in virtually all scientists. The researcher accepted that the results of an experiment existed based on the evidence that he sensed. He accepted the experience of sense data as adequate evidence of reality. He believed that if his senses transmitted reports of a thing that occupied time and space then that is evidence of its underlying reality. The thing is self sufficient, made of material matter. It is different from other things. But this is just an arbitrary mental conception.

These acceptances, beliefs, and assumptions are mental constructs. A more accurate reality is to accept that the mind-body and the world are one vast process out of which the mind constructs parts and labels the parts. But even that is not the foundation of reality. The foundation has been explored throughout this book.

Up to about 1920, classical physics assumed an objective existence of things and events the description of which everyone agreed upon. Later, physicists hypothesized a quantum mechanics subset of modern physics which holds out no objective existence. The view of those who believe in quantum mechanics is that the underlying reality depends on the point of a view of a conscious sensing human with a nervous system and a mind. The point of view could be altered with measuring instruments, by theory, by inference, by deduction, by logical induction, by assumptions, or by the observations of an experiment. The description of observed things or mathematically derived things depends on which theory the physicist believes, how he elaborates the sense data, and the mathematical description he employs.

Section 3. Patanjali taught how to live at the level of the underlying reality which he labeled, 'achieve unified knowledge of Godhead'

The teachings of Patanjali were adopted by the Vedanta Society. (Prabhavananda and Isherwood, 1953, p. 7- 20) Patanjali taught in detail how to achieve knowledge of Godhead which is the realization of the source of underlying reality. He taught how to realize enlightenment. These yoga aphorisms are thought to have been written between 400BC and 400AD.

Yoga practices, spiritual disciplines and techniques of meditation enable a person to achieve unified knowledge of Godhead. Godhead is defined as the reality which underlies this apparent ephemeral universe. Another definition of this achievement is "an effort to separate Atman, (the reality) from non-Atman (the apparent)".

The teachings use several words to describe entities of the human mind. These words reduce the complexity of the human environment. Reducing complexity is the process of ordering the perceptions in the human mind. Thus, these words are the beginning of decoding the apparent environment which human senses deliver to the mind.

The apparent environment is the expression of reality which has encoded information, action, Being, and entities into the world that humans experience.

The mind is called 'chitta.' It is made of three components:

1.) manas, the recoding faculty which receives impressions, sense data and information transmitted through the psychic field
2.) buddhi, the discriminative faculty which classifies impressions and reacts to them
3.) ahamkar, the ego sense which claims impressions for its own and stores them for knowledge which belongs to one human.

In Patanjali's system of decoding the underlying reality, God is another label for the underlying reality. It is claimed to be omnipresent. Reality is everywhere present in every living being and inanimate object. The God within the human is labeled, 'Atman' or 'Purusha.' Other labels and assumptions are as follows. The mind only appears to be intelligent and conscious. Mind is assumed to have a borrowed intelligence, borrowed from Atman which is identical to intelligence and consciousness. The mind merely reflects consciousness. Vritti is knowledge or perception; a thought wave in the mind. The mind is an instrument of knowledge. It is an object of human perception the same as sense data. In general, the Atman is the real seer but it is usually unknown to the mind.

Patanjali described the process of decoding the underlying reality. Every human perception arouses the ego sense which thinks, "I know this." Thus the ego is one of the streams of thinking but it is not the Atman, the Self. The ego sense arises from identification of the Atman with mind and senses. This is analogous to the light bulb.

Analogy of the electric light bulb

The light bulb is perceived as the electric current. One perceives electricity as a glass globe with wire filaments. It is true the electricity is partly within the bulb and also partly elsewhere. The electricity is analogous to the Atman. The bulb is analogous to the living human body. The filament is analogous to the ego sense. The filament identifies with the electricity. The bulb identifies with the filament. These identifications cause suffering such as the fear of loss of the filament or anxiety because of craving the electricity or the clinging to the bulb.

Patanjali clarified the limits of the ordinary mind

In the search for underlying reality, the mind is an instrument of knowledge. A single mind has limited ability to detect the communications from the underlying reality. Its reasoning is faulty. The intent to find the underlying reality is easily lost in the preoccupations with sense data. The mind can only imagine a few models of reality. There is wrong knowledge as well as right knowledge. There is verbal delusion and self deception such as believing that dreams are reality. Much of the imagination is based on words and the aggregate of words. One must remember, "Human language is like a drunken bear trying to beat out a tune on a broken drum while all the while the heart yearns to move the stars to pity."

Patanjali taught the method of gaining extremely difficult knowledge: samyana

Samyana is the combination of concentration, meditation and absorption. Concentration is difficult. It is difficult to separate sense data from the experience of sense data from the thoughts about the sense data. This lack of discrimination begins with the undifferentiated nervous system of the new born. A mature person can concentrate and also meditate on an object. Almost no one distills the Atman from the distractions and limitations of the human living process. Meditation on suitable objects results in absorption in Atman.

Section 4 The speculative nature of underlying reality constructed by physicists in the last 100 years

Some of the efforts of physicists in the last 100 years have been expended in research based on many unverified and un-falsifiable assumptions such as the objective reality of the physical world. A subset of physicists has discovered about 100 subatomic particles, more than 100 elements, and countless compounds of elements. Laws of chemical combinations have been framed. No process of combination of elements into living beings, consciousness or mind has been framed

In spite of the lack of understanding of the world, most scientists stand in awe of the complexity of living entities and life. They perceive beauty in the discoveries of science. They are loyal to the dogmas of science. They experience strong feelings about the achievements of the sciences. This is the way the contemporaries of Thales felt about the myths of creation. This is the way religious fanatics feel about the elements of their religions. The Ancient Chinese, and many other cultures, were also liable to be influenced by these perceptions, loyalties, feelings, and dogmas.

In science, a common question is "What is the unvarying principle?" Or "What is the prime mover?" Or "What is the first cause?" Or "What is the beginning of a chain of cause and effect?" Or "What is it that has zero entropy?" Many physicists believe in the model of reality that states that the underlying reality is the Higgs Boson, the God particle. They have convinced the many governments that, with the expenditure of about $10 billion US dollars and about ten thousand physicist years, the Higgs Boson will be observed at the Large Hadron Collider. These people are already convinced that the observation of the Higgs Boson will be the end of the search. The Higgs Boson could be interpreted as the definition of the underlying reality, the first cause. Now they claim to have detected the Higgs Boson; how can it be verified or falsified?

In the last centuries, how much deeper did the scientists probe into the underlying reality and its influences on human life?

Scientific search for the underlying reality called the Higgs Boson

An objective of scientific endeavors is to uncover the chain of cause and effect. When the causal connection is in doubt, probability mathematics often supplies a handle on the influences on an event. The scientific approach formally accepts layers of underlying reality or any other uncertain sources of reality.

Science is the current receiver of causal connections from the underlying reality. But the science organization labels the transmitter 'Nature,' not underlying reality. An irony is that the term, 'Mother

Nature' originated in primitive groups of people before Thales. An objective of science is to nullify the 'pagan' notion of a Mother Nature.

The science industry has invented several sources of creation. Some of the scientists in the last 200 years also invented theories of the origin of life, of the universe, and of matter. In the last 80 years, many physicists, and others, have subscribed to the latest theory of the origin of the universe, labeled, 'The Big Bang' a theory of creation of everything out of emptiness. Many physicists and others are convinced that the origin of matter is formulated in the dogma, 'The Standard Model.' The prestige of these scientists and the experimental success of the theory demonstrated by the atomic bomb and other evidence have persuaded sources of funding to fund the construction of large experimental apparatus to continue the research. Einstein stated that some physicist's objective was to read the mind of God. A large apparatus at FermiLab, which closed recently, was about seven kilometers in circumference. Another apparatus, called the Superconducting Super Collider was to be about 9,000 square kilometers in Texas but it was discovered to be unreasonably expensive to justify the research. The latest cathedral for the high priests of the science religion is called the Large Hadron Collider, about 23 square kilometers in Europe. To continue answering the question posed by Thales, these research monuments cost many billions of US dollars.

Consider the foundation of the Big Bang research. Absolutely nothing is known about the universe billions of years ago. There is no a way to falsify or verify that the universe had a beginning. If there were a beginning, there is no way to estimate how long ago that was. There are only guesses about what would have initiated the beginning. The assumptions begin when the conjectured universe was less than a second old. The Catholic Pope persuaded the scientists to promise to let God be the first cause. It is assumed that the universe was created out of nothing. There was a total void, no time, no space, no energy. There was a whim which triggered the creation. Out of nothing, emerged an entity at the temperature of about 10^{30} degrees Celsius. Space, time, particles, energy, black holes, particles grouped together to make elements, all instantaneously appeared. Then these entities gathered together to make the stars, galaxies and living things. These concepts are leaps of faith, not proven scientific truths.

Scientists and religious leaders study what is observed and invent structures of mental visions to bring order to the chaos. Only an infinitesimal amount of order has been codified. There is an unnamable suspicion that something is obscuring the vision such as dark matter, the speed of light, blurs, obfuscations and a lack of massive funding to bring order into the imagined structure. See *The God Particle: If the Universe is the Answer, What is the Question?* (Lederman, 2006)

One cannot avoid noticing the reflection of human tendencies toward power, wealth, material possessions, sensual pastimes, and fame in the fortresses of research organizations, in universities, in armed forces, in religions, and in industrial empires. In scientific researchers, one can decode these reflections in the lust for the Nobel prize, other prizes, exalted positions, high incomes, prestige, power, and goal seeking behavior. What is the percentage of scientific effort directed toward these reflections of human weaknesses? What is the percentage of scientific effort directed toward decoding the communications from underlying reality? How many scientists obey the communications from the enlightened spiritual leaders on how to think and act to enable one to decode underlying reality? An insignificant amount of effort is expended to discover all the powers of the human mind. What drove Thales to ask the questions?

This mental creation of the Standard Model and the Big Bang appear suspiciously like the myths that Thales was trying to avoid. Some scientists, convinced that the underlying reality is the answer to the question, "What are the ultimate building blocks of matter?"

Section 5 Research into reality as reported in European literature

Auerbach described 20 different interpretations of reality used in literature over about 2000 years in the European and Middle Eastern regions. There were many more that he did not write about, for example, the underlying realities built into the cultures in China and in the Americas. These cultures will be discussed below.(Auerbach, 2003)

Section 6 Reality in a sophisticated Aztec civilization before Europeans destroyed it

For the Aztec warrior in Mexico, in 1510 AD, death in battle, or better still death on the stone of sacrifice, was the promise of a happy eternity. Such a warrior who was killed in the field or on the altar was sure of becoming one of the "companions of the eagle," one of those who accompanied the sun from its rising to the zenith in a procession that blazed with light and was splendid with joy and then of being reincarnated as hummingbird, to live forever among the flowers.

The most important act of creation was the sun which was born of human sacrifice and blood. To keep the sun on its course so the darkness should not overwhelm the world, it was necessary to feed it every day with its food, the 'precious water,' human blood. Every time a priest cut out a human heart and held it up to the sun, the disaster that threatened the world was postponed for a day. Nothing was born or would endure without human blood sacrifice including the earth, rain, growth and nature. There were also warrior sacrifices. Women were sacrificed to the earth; their heads were chopped off while they danced. Children were sacrificed to the rain by drowning. Some of the blessed were fed a pain killer drug and thrown on the sacrifice to the fire. So humans were killed as sacrifices to feed all the many gods. Those sacrificed people accepted the event as an honor. (Soustelle, 1970)

A similar sacrificial ceremony has existed in Spain and in Mexico for hundreds of years. It is the sacrifice of a bull or a bullfighter in the ring for amusement.

Section 7. The Conviction of Reality Enabling European, Japanese and Russian human blood sacrifice in the last 100 years

One must put the Aztec blood sacrifice in the context with more recent sacrificial practices in war.

In World War I the British officers thought it was necessary for at least 10,000 men to die in an attack or else it was not a sincere sacrifice to the gods of National prestige and British manliness. A puny loss of life in battle would insult the gods of Victory, Fame, War and Honor. The British generals felt fulfilled when 57,000 British troops were killed or wounded in a single day. The sacrifice to the gods was 500,000 British in one year, 1916. The worship of National prestige, Manliness, Victory, Fame, War, and Honor gods was enthusiastic. The thrill of a woman hearing of the death of an entire British generation of young men as she sat safely at home in England was expressed, "...but the soul of the Empire will afford them. You shall break through with the cavalry of England for the greatest victory that history has ever known." (Hochschild, 2011, p. 214)

Americans traded about 100,000 Japanese lives for the lives of 500,000 American soldiers

Americans sacrificed about 100,000 Japanese civilians in about one second at Hiroshima. The atomic bombing was a measure to force the Japanese leaders to submit to the occupation of Japan by Americans and to preserve American soldiers. It was a triumph for the Science and Victory gods. It was celebrated all over the world.

The reality created by the Japanese to allow the murder of about 300,000 Chinese in Nanking

The signals sent from the Japanese military in China to the Japanese islands were filled with the noise a new god, "East Asian Co-prosperity Sphere." This was self deception by the Japanese which allowed the blood sacrifice of millions of Chinese civilians and Japanese soldiers.

The Japanese murdered 300,000 in Nanking China in about a week. This blood lust was transformed into worship of the war god by introducing noise into the information. The noise in the communication to Tokyo was that the" killing was necessary to practice killing skills." The war god required soldiers better trained in killing.

The Japanese justified the invasion of China in 1934, the destruction of the Chinese economy, and the murder of millions of Chinese over a duration of 11 years. It was the worship of the god of the East Asian Co-prosperity Sphere.

The Japanese considered it was an honor to die for the Emperor. This was similar to the Aztec desire to die in war.

For the Americans and the Germans, in contrast to the Aztecs, it was not considered an honor to die in their sacrificial rites which satisfied their reality.

The frequency of blood sacrifice proves it is an underlying reality of the human mental archetype.

Section 8. The reality that was created to allow the genocide of original Americans, Africans, and Australians.

Let us note the noise in the decoding of the signals from the Europeans when they were worshiping the Humanity and Progress gods from about 1550 to 1950.

The signals sent from the mass murderers to the leadership of the European countries encoded the 400 years of murderous genocide in the Americas, Africa, and Australia. The noise, "freeing the land from the savages" covered up the savage destruction of several civilizations. The murderers introduced another noise, "the denial" of the vast genocide being undertaken by Europeans in North and South American, Africa and Australia.

When the noise is removed, the information remaining is that the Europeans murdered 10 to 20 million of the original owners of the land and destroyed hundreds of cultures, languages, and whole civilizations.

After introducing noise to deceive themselves about the fact of 400 years of genocide and massive destruction of several civilizations, the Europeans masked the same genocide and massive destruction of cities, cultures, and languages behind noise. These sciences were perfected and applied to European victims.

After murdering entire civilizations in Africa, Australia North America and South America, Europeans applied their mass production of death to other Europeans.

The Russians and the Americans joined the Europeans in two World Wars.

They invented noise to mask the unimaginable evil of their blood sacrifice to the gods of National prestige, Manliness, Victory, Fame, War, and Honor.

Section 9. The reality created by the Europeans and Russians to allow them to murder 20 to 50 million people in concentration camps and forced labor projects

The Americans and Europeans believed in the false gods of Humanity and Progress, and in the Goodness of Man.

The Europeans, Russians, and Americans applied their mass production of death to other Europeans. In addition to the 100 million extra dead during the wars, tens of millions of Europeans, Russians and Americans were trapped in German and Russian concentration camps from about 1940 to about 1990.

Several million people, mainly Jews, were sacrificed by the Germans. This was considered necessary for the German god, "Purity of the gene pool." This was noise and self-deception, in the communication.

When the victims arrived at the camps, they were deceived by the noise: their belief in the false Gods of Humanity and Progress, and in the Goodness of Man. The jailers and also the inmates believed in these gods.

The criminally insane conditions they invented were beyond most people's ability to perceive or believe. Therefore, the victims denied the reality. Most of those who entered the camps denied that the horror existed. Those died right away.

The reality the victims had created in their minds was based on past experience. The reality of the Jews before the concentration camp was the worship of Humanity and Progress. The first days in camp were unreal and were often identified as nightmares. Prisoners had to convince themselves this was real. To the extent that reality is a mental or cultural construct, the camps were unreal. They had been convinced that there had been 200 years of advance in well-being. They praised the false gods, of Humanity and Progress which had taken the place of the Jewish God. The new age had arrived with faith in humanity replacing the faith in God. How did the prisoners make sense of the inhumanity massing to destroy them? Evil on this scale was not believable. The dream of Hell, which for millennia had haunted human consciousness, was then actual reality. Some people believed they would wake up and it would disappear.

By coming to admit the actual evil of the European world they were in, they gained a perspective emerged: selfhood, realism and the desire to outlive the evil. (Des Pres, 1976, pp. 83- 87)

When the concentration camp prisoners removed the noise in the message, then the victims mentally invented a new reality. They had to cooperate with each other to live, to learn, to resist, and to fight back. They were driven to write the facts of the evil hiding within human nature. They were desperate to ensure that some of them would survive to remove the noise in the belief system of the false gods for the remainder of humankind. They lived for the remainder of humankind; to save it from creating more Hell. (Des Pres, 1976)

The Russians copied the Europeans by murdering tens of millions of Russians. After World War II, Stalin and the other the jealous leaders in the USSR were convinced it was necessary to starve to death about 10 million to preserve Stalin's position in power. The USSR government was convinced it was necessary to kill about 20 million in camps used for forced labor to make an industrialized country.

Section 10. The Europeans introduced noise into the information about the reality of centuries filled with their destruction of civilizations and their theft of four continents

Let us note the noise in the decoding of the signals from the Europeans when they were worshiping gods of White Man's Superiority, Humanity and Progress. The signals sent to the leaders of the European countries from the gangs of criminals encoded noise into the information. One noise was "freeing the land from the savages." Another noise was the "denial" of the vast genocide being undertaken by Europeans in Africa, North and South American, and Australia. Another noise was "conquest, civilizing the sub-human savages, and righting the wrongs against our noble adventurers."

After one removes the noise, the information that remains is murder of 10 to 30 million of the original owners of the land, and destruction of hundreds of cultures, languages, and whole civilizations. The noise, "denial," was the covering up of the Hell that the Europeans had brought to four continents.

Section 11. The reality created by the British to allow them to steal several cities from the Chinese and enforce opium addiction

The British methods of destroying entire civilizations and cultures were initiated into China. The British confused the facts with noise, "Aggressive Chinese" and "refusing to buy opium but destroying it instead."

The British signaled noisy concept to England that they had put a stop to the aggressive Chinese who wanted the British to get out of China. They used the destruction of opium and other minor acts as reasons to attack with military and navel war. The British began to destroy cities until the Chinese agreed to buy unlimited opium, to give Britain several port cities, and to allow the destruction of the moral foundation of China.

The British interjected the noise, "they had subdued primitive natives and had brought civilization to the pagans." Another noise was that "the land was inhabited by sub humans." Therefore it could legally be claimed for the glory of Queen Victoria.

Section 12. A phenomenon which communicates an underlying reality within the human mental archetype: the need for war

War is communicated to humankind as history, as autobiography, as travel writing, as narratives by those who were there but quite different from these classes of writing. Much of war is presented in delusions conjured up by those who were not there in the mud with a missing leg having just killed another man. They glorify or ignore the murder of 200 million young men in the last 100 years, the future of humankind turned into ground meat. They ignore the 500 million men who were mentally destroyed and will never enter what others call reality. The grandest and most noble lie was, "This is the war to end all war." (Hynes, 1997)(Fyfe, 1988)

Section 13. Pretence that an idealistic mental conjecture such as socialism, communism or a state based on welfare is functioning

**according to the idealism but simultaneously denying
the failure of society to provide the basic necessities of human life**

There are several sources of noise in this hypocrisy.

1) The denial by a large group is a type of noise introduced into the facts and sense data. This process of group denial of apparent reality is not unusual.
2) The idealism deceives those fanatics who believe the lie and also deceives others who oppose the idealism as delusion that violates fundamental laws of economics.
3) There is a noise that is an implied addiction by most people that there are many elements of society that do not supply the basic necessities.
4) There is the noise which covers the extreme selfishness of those who are benefiting from the forced implementation of the idealism.

Schwartz observed a type of thinking by a group that obstructs scientific adaptation. He was witnessing the forced implementation of communism with Chinese characteristics. This type of thinking is pervasive in many large groups such as whole nations. An example of this kind of thinking in the form of a riddle: most humans are bound by culture, by their short life in history, by social position within social levels, by the limitations of the human mind, and other forces that retard adaptation in the scientific evolution sense. These bounds could be called apparent reality.

Opposed to this apparent reality, most humans believe their behavior, their expectations of future world, are actually based on their individual beliefs or their group convictions. They ignore apparent reality. This opposition group is composed of delusional people pretending that in all cases that their mentally invented world of how life ought to be has replaced apparent reality. Tiny changes toward implementing this delusion are touted as proof the new has arrived. This can be seen in communist, socialist, American Democratic political party, and other idealistic groups. How can the opposition group ignore observable facts? How does one answer this riddle? (Schwartz, 1984, p. 6-10)

Another example of this mass delusion is half a society that believes in the ability to create goods, services, wealth even though the observable facts are that people are inherently sinful. The invisible hand of the capital market makes this creation happen. This is the capitalist group.

The other half of society opposes capitalism but believes that because people are sinful there must be a government that controls goods, services, and wealth in spite of the fact that governments are composed of sinful people. Also no government has ever been able to produce the necessities without the aid of the capitalists. They ignore the historically observable facts.

Many people deny apparent reality. These people claim that their "truth" is reality and in addition, their "claims of truth" are reality. They also claim their "truths" do not contradict apparent reality. They claim their truths are not based on wishful thinking, nor based on their desires for their own individual future benefit, nor based on clinging to their personal goods and interests. (Schwartz, 1985, p. 6)

There was a recent observation of resistance to change because of personal desire for one person's benefit. In China, public argument has the ancient flavor of vicious attacks intending to get the attention of more people much like a taxi driver screaming in the street. Even among university professors the attacks were venomous, making personally insulting accusations based on quotes that were taken out of context. The professors appear as selfish, defending their careers but not trying to teach students better. They are hypocritical and weak spirited. The spirit of dispassionate disagreement with clearly stated problems. They fight threats of change fearing a loss of professorship. This was noted in 2011. (Zha, 2011)

Confucius denied apparent reality. He transmitted the way of the ancient gentlemen from hundreds of years before him who he believed lived by the following code. The code could be used as noise to confuse one's intention and hide behind lofty ideals.

Chih, uprightness and integrity.

Yi, righteousness, doing the right thing.

Chung, loyalty and consideration of the feelings and needs of other people.

Chu, do not do onto others what you do not want done to yourself.

Jen, loving-kindness, empathy, inner intuition leading to sympathy for other living beings.

Li: In Confucius' era, this principle was widely interpreted to mean different ways of behaving. At first the '*li*' was a vessel used in sacrifice ceremonies. It evolved into a manner of conduct, tending to formalize manners and etiquette.

Since Confucius pursued his ambition to become a high ranking dignitary, he learned and taught how to behave with correctness. He endowed *li* with more importance than just knowing when to bow. He perceived it as comprehending the totality of man as a social being. It meant utter integrity, passionate striving after wisdom, undaunted belief in courage, dispensing justice, and feeling compassion.

Ever since Confucius taught *li* in Chinese culture, it was practiced as deference to a person superior in rank, son to father, official to emperor, wife to husband, younger brother to elder brother, woman to man. Li has had a profound influence in all history.

It was an expression of sincerity. It was more than a display of silk and jade. Symptoms of true *li* are restrained actions, moderate judgment, and persistent dignity.

The noise of 'Golden Age' was based on wishful thinking. This was another vague noise used to hoist oneself to a higher idealism be referring to perfect men long forgotten in history. This conviction, that long ago in the past Golden Age the kings were virtuous, continued until today in Chinese culture.

Later, Mencius and others codified laws of reward and punishment that recognized the reality that people often violated these assumptions about human natural virtue. This was closer to apparent reality. Later, Hsün-tzu codified laws that recognized the reality that officials representing the emperor were the worst violators. Later proposals recognized the reality that the emperor was usually the source of disorder in society and the cause natural disasters. To this day, no one could convince the emperor or the holders of the top power in the Chinese Communist Party to obey laws. The denial of apparent reality has continued for several thousand years.

These recognitions of levels of reality from long before Confucius until today are examples of massive conviction of imaginary wishful thinking or delusional reality.

Another example of the belief in an impossible production of food without work or money. Half of Chinese society believes in the ability to create goods, services, wealth even though the people are inherently sinful. The invisible hand of the free capitalist market makes this creation happen.

The other half of society believes in the delusion that because people are sinful there must be a government that controls goods, services, and wealth in spite of the fact that governments are composed of sinful people. This was the communist ideal of "more, better, faster." The delusion that was sent from Mao was that people would work hard for no pay and the harvest of food would be enormous. More astonishing is that the record shows that no government has ever organized the productive resources to produce a profit.

The Chinese communist party also claimed their "truths" do not contradict apparent reality such as economics and agriculture, steel making and farming. It claims its truths are not based on wishful thinking, nor based on their desires for their own individual future benefit, nor based on clinging to their personal goods and interests. (Schwartz, 1985, p. 6)

In fact, the new world order is indeed based on clinging, individual benefit, and the interests of each individual man in spite of the damage to society as a whole. The next section addresses Mao clinging to prestige and craving a god like figure of himself in spite of the death of about 20 million Chinese.

Section 14. Examples of a massive denial of fundamental laws of nature in farming and in steel production

How could millions of Chinese people deny economic laws such as supply and demand and the automatic function of the capitalist free market to produce more goods. How could 100 million farmers ignore the laws of planting and harvesting; the obvious laws of nature.

In spite of the laws of economics, agriculture, steel making and farming and in spite of the laws of science, often these fixed false beliefs based in communism were obeyed as though they were laws of nature.

The hope was that because Mao said impossible methods of farming and steelmaking were possible, then it was true. Communist commands to destroy the landlords and to convert individual farms into communal farms were obeyed. This caused temporary changes such as taxing the wealthy to provide housing for those who refuse to work. But these changes cannot be permanent because the changes violate the laws of the state or the laws of science or the laws of economics. Those who refuse to work reduce the benefits of the whole group until the wealthy cannot afford to pay enough taxes to support the wastrels. This is one of the results of communism and, incidentally the American welfare state.

The delusional thinking of communal farms was that people would be happy to work all day without pay to get free food. But the people refused to work much since they were promised plenty of free food. There is a tendency for people to work hard if they get to keep the profit of the work, a capitalist motive. Otherwise, they work slowly and take naps. So there was a tiny harvest and about 22 millions of people starved to death. The number of deaths was based on the actual population for 1958 and 1962 compared against the population with an historical 2% per year increase. The starvation was evidence that communist farming based on wishful thinking and promises were just noise.

An example of denial of reality (noise) by hundreds of millions of people: during a few years of communist ideology in China follows. Mao Zedong decided what the grain production would be. Mao Zedong introduced noise that the farmers could use communist thinking, not the methods they had learned in thousands of years of experience, to increase the harvest. He ordered food production to be increased by an unreasonable amount based on the noise that communist ideas would work. He did not consult with farmers about what was possible. Then the people were required to deliver grain to the government based on impossible imaginary numbers based on lies.

The farmers introduced the noise of denial. They made impossible promises of food production. They lied about the quantity of the harvest. They hoped the noise would hide the reality of low food production per hectare that their ancestors had proven for thousands of years. So the people had to deliver the grain to the Government instead of eating it.

Mao Zedong ordered that all birds were to be killed because they eat grains and there was not adequate food for people. People killed all the birds. Because of no birds, insects became a menace. Mao Zedong ordered too much insecticide to kill all insects. The result was there are no bees in China today so the crops must be pollinated by humans. The Chinese starved to death by the tens of millions.

This shows the importance of removing the noise from the accurate information.

There is another example of hundreds of millions of people infected by a massive delusion of reality. Mao Zedong ordered everyone to make steel to achieve an impossible steel production. The noise was that farmers could make steel. Making steel requires iron ore, a refining industry and other elements. Farmers do not know how to make steel. However, about 100 million farmers stopped making food and even stopped harvesting the available food. They melted anything made of metal instead of farming. This caused a severe shortage of food which resulted in starvation. They cut down huge forests for fuel to melt metal objects that already existed. They did not make new steel; they melted whatever metals they had. Then they pretended that would count as steel. The metallic substance was not steel; it was a mixture of metals which was worthless for any use. They kept the melting ovens going 24 hours a day seven days a week for several years.

The cutting of trees for steel reduced many areas to desert which have never again been planted with trees. The bare land was blown in dust storms into Beijing and other places. The rains washed the soil into the rivers and into floods. The floods ruined towns and also the farmland.

A result of this noise of denial by millions of people was almost no food for the farmers, substantially less forests, famine which killed about 60 million people but produced no new steel. Other industries were neglected. School children were not educated. Employees were devoted to cutting trees and melting metal. Babies were left with no care and no breast milk. Air was polluted by oven smoke. Cities were polluted by dust. Farmland and towns were flooded.

The people introduced the noise of lies; claiming huge steel production in addition to claiming impossible food production. Required to lie, the people lost respect for the government. Those who spoke about or wrote about the failure of the whole civilization were put to death. Finally the steel and the food goals were cancelled.

The apparent reality changed a little because there was a vast amount of worthless metal chunks. So the Communist Party leaders (analogous to other group belief systems) were encouraged to push society toward another inherently impossible belief system. The Chinese Communist Party claimed a huge steel production.

The population denied the famine, the loss of forests, and the fact that the melted objects could not be used as steel. Many years later

Mao Zedong had to admit the whole process failed to make steel. No one admitted the process failed financially, failed to raise worthwhile children, destroyed the forests, destroyed the bees, and undermined the belief in communism. These systems of massive belief in the impossible, was the noise of self deception. Communism, and the noise it uses to mask failure, continues in various places to this day. This has been the history of denying apparent reality in favor of dreamy mental ideals.

No one was permitted to talk about how the processes failed financially, failed to raise worthwhile children, destroyed the forests, destroyed the bees, and undermined the belief in communism.

Section 15. Several books that provide insight into the range of realities under which people perceive their lives

The following is a suggested set of books on reality. The intention is to introduce a wide range of notions about reality. Clearly, reality is not distinct. However, many people have attempted to define it.

Juan Gabriel Vasquez described in detail people noticing the apparent reality that restricted their ambitions in *The Sound of Things Falling*. (Vasquez, 2013)

Hlatky, Stefan and Booth, Philip (1999). *Understanding reality: a commonsense theory of the original cause*, Charlbury, Oxfordshire: Jon Carpenter

Zhang, Boduan,[active 10th century-11th century] (1987). *Understanding reality: a Taoist alchemical classic*, Chang Po-tuan, ed., concise commentary by Liu I-ming; Thomas Cleary, trans. Honolulu, HI: University of Hawaii Press.

Davies, Paul Charles William, and Gribbin, John (1992). *The matter myth: dramatic discoveries that challenge our understanding of physical reality*, New York: Simon & Schuster.

Burbidge, John W. (2013). *Ideas, concepts, and reality*, Montréal: McGill-Queen's University Press.

Katz, Jerrold J. (1971). *The underlying reality of language and its philosophical import*, New York: Harper & Row.

Kosso, Peter, (1998). *Appearance and Reality: An Introduction to the Philosophy of Physics*, New York: Oxford Univ. Press.

Lederman, Leon (2006). *The God Particle: If the Universe is the Answer, What is the Question?* New York: Mariner.

Libet, Benjamin "A Testable Field Theory of Mind-Brain Interaction." *Journal of Consciousness Studies,* Summer 1994, No.1: 119.

Section 16. Mental disorders and Failures of mind

Multiple authors (2013). *Diagnostic and statistical manual of mental disorders, DSM-V,* 5th ed., Washington, D.C.: American Psychiatric Association.

This catalog of mental illness is included to show the broad range of what some people consider to be distorted or disordered point of view. But the patients are convinced that they perceive the true reality of the world. In this case, the medical staff has to destroy the convictions of reality.

Description

Mental illness, personality disorders, psychosis, in addition to criminal intent, ordinary lying, cheating and stealing; all of these obstruct the channels to the underlying reality. Some of these obstructions are standardized in this DSM-V.

This is the standard reference for clinical practice in the mental health field. Used by clinicians and researchers to diagnose and classify mental disorders. Since a complete description of the underlying pathological processes is not possible for most mental disorders, it is important to emphasize that the current diagnostic criteria are the best available description of how mental disorders are expressed and can be recognized by trained clinicians including:

representation of developmental issues related to diagnosis;
integration of scientific findings from the latest research in genetics and neuro-imaging;
consolidation of autistic disorder, Asperger's disorder, and pervasive developmental disorder into autism spectrum disorder;
streamlined classification of bipolar and depressive disorders;
restructuring of substance use disorders for consistency and clarity;
enhanced specificity for major and mild neuro-cognitive disorders;
transition in conceptualizing personality disorders;
new disorders and features.

The Table of Contents of DSM-V lists the main types of abnormal reality

Cautionary Statement for Forensic Use of DSM-V (This adds the legal obstruction of reality to the mental obstructions.)
Section II Diagnostic Criteria and Codes
Neurodevelopmental Disorders
Schizophrenia Spectrum and Other Psychotic Disorders
Bipolar and Related Disorders
Depressive Disorders
Anxiety Disorders
Obsessive-Compulsive and Related Disorders
Trauma- and Stressor-Related Disorders
Dissociative Disorders
Somatic Symptom and Related Disorders
Feeding and Eating Disorders
Elimination Disorders
Sleep-Wake Disorders
Sexual Dysfunctions
Gender Dysphoria
Disruptive, Impulse-Control, and Conduct Disorders
Substance-Related and Addictive Disorders
Neurocognitive Disorders
Personality Disorders
Paraphilic Disorders
Other Mental Disorders
Medication-Induced Movement Disorders and Other Adverse Effects of Medication
Other Conditions That May Be a Focus of Clinical Attention

Acocella provided a non-technical description of a personality disorder listed in the DSM-V, narcissism. It gives a clear meaning to the concept of disorder. (Acocella, 2104)

Section 17. Virtual reality; voluntary shared delusion

Multiple authors (2004). *Understanding reality television*, Su Holmes and Deborah Jermyn eds., New York: Routledge.

Table of contents of *Understanding reality television*:

Holmes, Su and Jermyn, Deborah "Introduction: understanding Reality TV"

Clissold, Bradley D., "Candid Camera and the origins of Reality TV: context of an historical precedent"

Jermyn, Deborah "This is about real people!: video technologies, actuality and affect in the television crime appeal"

Biltereyst, Daniel, "Reality TV, troublesome pictures and panics: reappraising the public controversy around Reality TV in Europe"

Stephens, Rebecca L., "Socially soothing stories? Gender, race and class in TLC's A Wedding Story and A Baby Story"

Sherman, William R. and Craig, Alan B. (2003). *Understanding virtual reality: interface, application, and design*, San Francisco, CA: Morgan Kaufmann

Table of contents of *Understanding virtual reality*

Interface to the Virtual World--Input

Interface to the Virtual World—Output

Rendering the Virtual World

Interacting with the Virtual World

The Virtual Reality Experience

The Future of Virtual Reality

Multiple Authors, *Metaphysical grounding: understanding the structure of reality*, Fabrice Correia, Benjamin Schnieder eds.

Summary of *Metaphysical grounding*:

Some of the most eminent and enduring philosophical questions concern matters of priority: what is prior to what? Is matter prior to mind?

How such questions have to be understood? Can the relevant notion of priority be spelled out? And how do they relate to other metaphysical notions, such as modality, truth-making or essence?

This volume of new essays in contemporary metaphysics, addresses the metaphysical idea that certain facts are grounded in other facts.

Table of Contents of *Metaphysical grounding*:

1. Fine, Kit "Guide to ground"
2. Daly, Chris, "Scepticism about grounding"

3. Audi, Paul, "A clarification and defense of the notion of grounding"
5. Della Rocca, Michael, "Violations of the principle of sufficient reason (in Leibniz and Spinoza)"
6. Williams, Robbie, "Requirements on reality"
8. Lowe, E, J., "Asymmetrical dependence in individuation"
9. Azzouni, Jody, "Simple metaphysics and 'ontological dependence"
10. Liggins, David, "Truthmakers and dependence"

Dyke, Daniel (1628). *The mystery of selfe-deceiuing. Or A discourse and discouerie of the deceitfulnesse of mans heart. Written by the late faithfull minister of Gods Word Danyel Dyke, Batchelour in Diuinitie*, Published since his death, by his brother I.D. minister of Gods word, London: Printed by William Stansby, and are to be sold by Michaell Sparke, in little Old-bayley in Greene-arbour, at the signe of the Blew Bible.

Varki, Ajit and Brower, Danny (2013). *Denial: self-deception, false beliefs, and the origins of the human mind*, New York: Twelve.

Whaley, Barton (2006). *Detecting deception: a bibliography of counter-deception across cultures and disciplines*, 2nd ed., Washington, DC: Office of the Director of National Intelligence, National Intelligence Council, Foreign Denial and Deception Committee.

Multiple authors (1999). *Understanding representation in the cognitive sciences: does representation need reality?*, Alexander Riegler, and Markus Peschl, and Astrid von Stein, eds. New York: Kluwer Academic/ Plenum Publishers.
Table contents *of Understanding representation:*
Peschl, Markus F. and Riegler, Alexander, "Does Representation Need Reality"
Dorffner, Georg, "Different Facets of Representation"
Alfredo Pereira, Jr., "Representation in Cognitive Neuroscience"
Hutto, Daniel D., "Cognition without Representation?"
Robinson, William, "Representation and Cognitive Explanation"

Haselager, Pim, "Neurodynamics and the Revival of Associationism in Cognitive Science"

Bressler, Steven L., "The Dynamic Manifestation of Cognitive Structures in the Cerebral Cortex"

Mogi, Ken "Response Selectivity, Neuron Doctrine, and Mach's Principle in Perception"

von Stein, Astrid. "Does the Brain Represent the World? Evidence Against the Mapping Assumption"

Chandler, N., Balendran, V., Evett, L., Sivayoganathan, K., "Reality: A Prerequisite to Meaningful Representation"

Gardenfors, Peter, "Does Semantics Need Reality?"

Weiss, S., Muller, H. M., Rappelsberger, P., "Processing Concepts and Scenarios: Electrophysiological Findings on Language Representation"

Wallin, Annika, "Can a Constructivist Distinguish between Experience and Representation?"

Rothman, Milton A. (1992). *The science gap: dispelling the myths and understanding the reality of science*, Buffalo, NY: Prometheus Books.

Subbotskiï, Eugene. V. (1992). *Foundations of the mind: children's understanding of reality*, New York: Harvester Wheatsheaf.

Rihani, Samir (2002). *Complex systems theory and development practice: understanding non-linear realities*, New York: Zed Books.
Table of Contents of *Complex systems:*

Rise and Fall of Complexity
Certain Unpredictability
Cooperation and Competition
Linear Recipes for a Complex World

Section 18. Perception of the psychic field

DeStefano, Anthony (2011). *The invisible world: understanding angels, demons, and the spiritual realities that surround us*, New York: Doubleday.

Summary of *The invisible world:*

There is a kind of ESP "a haunt detector." It detects whenever something mysterious or supernatural has occurred. This book explains the reality of the spiritual dimension that surrounds us and shows how it is immediately accessible to everyone. In this book, all aspects of the spiritual realm are discussed, including the existence of angels and demons.

Howard developed the scientific description of psychic phenomena including the following concepts and the math descriptions.

1) The Geometry of Precognition
2) Time to Formulate the Laws and Hypotheses of Psychic Science
[3] Discovering the Hypotheses of Psychic Science
5) Can an Ordinary Person Be Trained to Use the Psychic Senses?
(Howard, 2012e)

Section 19. Reality encoded in logic

Heidegger, Martin (2010). *Logic: the question of truth*, Thomas Sheehan, trans. Bloomington, IN: Indiana University Press.

Table of Contents of *Logic:*

The first, most literal meaning of the word "logic."
A first indication of the concept of the subject matter.
The possibility and the being of truth in general; Skepticism.
The place of truth, and logos. (proposition)
The basic structure of logos and the phenomenon of making sense.
The conditions of the possibility of logos being false; the question of truth.

The presupposition for Aristotle's interpretation of truth as the authentic determination of being.

The idea of a phenomenological chronology.

The conditions of the possibility of falsehood within the horizon of the analysis of existence.

Concern-for and concern-about, authenticity and inauthenticity.

Preparatory considerations toward attaining an original understanding of time; a return to the history of the philosophical interpretation of the concept of time.

The influence of Aristotle on Hegel's and Bergson's interpretation of time.

A preliminary look at the meaning of time in Kant's Critique of pure reason.

The interpretation of time in the Transcendental Analytic.

The function of time in the Transcendental Logic;

The question of the unity of nature, transcendental unity of apperception.

Time as the universal a priori form of all appearances.

Time as original pure self-affection.

The question about the connection between time as original self-affection and the "I think."

Interpretation of the First Analogy of Experience in the light of our interpretation of time.

The "now-structure" that we have attained: its character of referral and of making present.

The phenomenal demonstrability and limits of Kant's interpretation of time.

Time as an existential of human existence: temporality and the structure of care.

Section 20. A Buddhist reality

Hundreds of religions have separate realities. Several were presented briefly throughout the book.

van Gorkom, Nina (1990). *Understanding reality*, London: Dhamma Study and Propagation Foundation.

Buddha (500BC). "10 Satipatthana Sutta The Foundations of Mindfulness," *The Middle Length Discourses of the Buddha: A New Translation of the Majjhima Nikaya*, Bhikkhu Nanamoli and Bhikkhu Bodhi, trans., Boston: Wisdom Pub.

Dalai Lama (2012). *From Here to Enlightenment: An Introduction to Tsong-Kha-Pa's Classic Text the Great Treatise on the Stages of the Path to Enlightenment*, Boston: Snow Lion

Dhammajothi, Mădavacciyē, Himi and Dhammajothi, Medawachchiye Thero (2009). *Concept of Emptiness in Pali Literature*, Colombo: International Publishers.

Appendix D: An Attempt to Label Some of the Separate Levels of Mind and Consciousness

Abstract

It is possible for some people to realize higher levels or more valuable states of mind and consciousness than other people. The Buddha provided mental tools that help to define and interpret communications from the underlying reality.

It is possible for some people to realize higher levels or more valuable states of mind and consciousness than other people

The literature of consciousness and mind in relation to the brain and body are fragmented; not organized into a comprehensive network of inter-related mental constructs. Thus, the many sciences bearing on these entities have not achieved a comprehensive description of the whole.

Neither the math nor the mental tools have matured adequately to formulate all the conjectures or hypotheses necessary to describe the whole field of mind, consciousness, and body. A short set of conjectures about body, brain, consciousness and mind follow for the purpose of stimulating scientific investigation. The objective is to form a focused taxonomy of concepts to organize the body, brain, consciousness and mind. This may stimulate people to add new ideas so that a complete science of brain, consciousness, body, and mind science will be defined by the classification of the ideas. (Howard, 2012a, Chapters 9, 10, 11)

Hypothesis: The consciousness and mind are generated by the changing processes of the physical brain and body.

Hypothesis: Reciprocally, many changes in the physical brain and body are caused by the conscious mind.

Hypothesis: the conscious mind is an emergent system based on the entire physical substrate, especially the totality of all information transmissions.

The Buddha provided mental tools which help to define and interpret communications from the underlying reality

These tools enabled a man to rise to a higher level of mind. The Dalai Lama recited the Buddhist teaching on emptiness as a mental foundation for escaping from the troubles of ordinary life. There are two kinds of emptiness in Buddhist philosophy. See Appendix C Section 1. The Buddha taught the detailed method of perceiving reality.

One kind is "inherent objectivity" in things that exist in the system of human perception based on an undistorted mind.

Another kind of emptiness rejects the notion of things existing objectively. All phenomena are dependently originated. Everything exists because of other things. The representation of such phenomena in the human nervous system and minds depend on language, agreed definitions, and learned taxonomy, the allowable scientific objects of study. Because of this representation, living beings identify individual items and events that they use to give order the interconnected world where everything depends on other things to exist. After one has removed the assumption of inherently existing things by purification of the mind, there is no ground, no self supporting basis, nothing to grasp or cling to.

Although this kind of emptiness with no intrinsic reality, is based on the experience of the few practitioners who are enlightened, it includes the fact that people correlate their perceptions with each other. This correlation is adequate to convince most people that there is intrinsic reality. However, as a man evolves toward the enlightened state, he rejects his conviction of intrinsic existence. He manifests a higher level of mind.

Emptiness based in physics

Physical scientists claim they have attained a higher level of mind due to their enlightened point of view. The physical science world view has a third kind of emptiness. Quantum mechanics asserts no objective existence. The underlying reality depends on the point of view of a consciousness embedded in a human sensing the phenomenon. The human senses are augmented with a nervous system and a mind. (Dalai Lama, 2012, p.143-144)

The classical physical science world view assumed an objective existence of whatever the humans senses delivered to the nervous system.

The Buddha taught methods to achieve, to realize in one's mental life, higher levels of mind. After realizing these levels, the monk experiences different ordinary mental levels. These mental levels allow enhanced interception of communications from the underlying reality.

A proposed set of mind levels, a few of which were taught by Buddha (Buddha, 500BC)

Level 0: Non-living, or post living matter; unorganized brain and body, atomic and molecular matter and their properties and behavior. This is physical matter, physical mechanisms, changing in accord with electric and chemical reactions.

Level 1: Living Brain-Body cells: The discrete and systematic functions of individual living cells. There are mathematical descriptions and chemical descriptions of living processes. Some of the physical symptoms of life follow:

neurons, ligand-receptor pairs, and hormones.

ATP, citric acid cycle etc.

Reactions to body chemistry

Reactions to body generation of electromagnetic phenomena

Level 2: The operations of the physical component organs:

Normal organ functions, input chemicals, information, and the output of wastge

Disorders: Malnutrition, lack of muscular vitality, trauma, disease, and lack of functioning organs that can stop the activation of the body, brain, and nervous system.

Level 3: Autonomic nervous system operations, not necessarily registering in the mind or consciousness.

Reflex actions of body.

Habitual actions of body.

Action of autonomic nervous system.

Conscious operations of autonomic nervous system under willpower.

Level 4: Mutual generation of brain-body and mind.

What is mind? Aristotle (c. 630BC) defined its existence in combination with body. "Soul (mind) and body react sympathetically upon each other; a change in the shape of the soul produces a change in the shape of the body and conversely." (Aristotle, 1936)

This is a conjecture, a hypothesis that can be accepted or subjected to experimental verification or falsification.

Hypothesis: The consciousness and mind are generated by the changing processes of the physical brain and body. Reciprocally, many changes in the physical brain and body are caused by the consciousness and mind.

The natural order has many examples of one phenomenon being regenerated by another which in turn regenerates the first phenomenon. An example is electricity and magnetism which have been thoroughly studied and applied. (Adair, 1987, Chapter 8) Another example relevant to consciousness with a detailed preparation to the concept, is given in *Decoding Reality: The Universe as Quantum Information.* (Vedral, 2010, p. 215-216)

Level 5: Brain Physical substrate

The physical basis of mind and consciousness has several levels. These are the domain of neuroscience. (Stein and Stoodley, 2006, p. 111)

Feeling induced brain-body responses

Electromagnetic generation of complex brain field, EEG, and nervous system field, EKG.

Language

Memory

Components of seeing, attaching meaning, perceiving, with mind

Components of hearing, attaching meaning, perceiving, with mind

Components of smelling, attaching meaning, perceiving, with mind

Components of tasting, attaching meaning, perceiving, with mind

Components of touching, enteroceptors, exteroceptors, proprioceptors, etc skin senses, abdomen signals.

Other brain functions that have been identified in science fields to date.

Part of meaning is identification of sense data from body, or from World1 outside body, or from mental inventions in World3.

Level 6: Birth Brain

Pre birth development of brain.

Pre-birth development of instincts, collective consciousness, knowledge of communication with mother, breast reaction

and other elements contained within brain and nervous system.
Unconditioned reflexive thought.

Level 7: Non conscious mind operations
Reflex reactions to external stimuli such as Pavlov's dogs
Physical body operations not registering in the mind.
Creation of mental objects such as dream stories.
Creation of disordered and psychotic mental constructs.
Destruction of mental constructs, memories, feelings.
Destruction of physical abilities, speech abilities.
Archetypes
Instincts

Level 8: Consciousness Function; Communication System of Information between brain, body and mind-information processing system not including the information itself

[This consists of information channels and the combined electromagnetic field generated by the entire body.]

Information diffusion between psyche mind and physical
body and the reverse. (This has been described mathematically by Howard, 2012c)

Information diffusion between non-conscious and conscious mind.
Ligand-receptor, hormone, enzyme etc. information systems.
Physiological information systems: heartbeat, temperature,
abdomen sensations.
Homeostatic information system.
Extra Sensory Perception [Various types of information
diffusion leading to conscious attention or non-conscious
acquisition of information not employing senses(ESP)].

Level 9: Functioning Mind:
[Mind is an emergent entity which has manifested out of the physical body. It is a non-physical process, existing only as the emerging state from the brain, nervous system, body, and electromagnetic fields. It does not exist unless it is transmitting, receiving and processing information. It needs a trigger to function.]
Function of creating information.
Function of encoding information using a coding key that is understood by the intended receiver.
Function of initiating transmission of information to intended
targets of the physical brain and body, including speech

organs and muscular activation.

Function of a receiver of information from the brain-body.

Function of decoding information.

Hypothesis: the conscious mind is an emergent system based on the entire physical substrate, especially the totality of all information transmissions.

The emergent system is a decentralized, massively parallel distributed processing system.

The conscious psyche, although it is based on the physical body, is an information processing system which is not physical and does not operate on the laws of the physical world. It has its own laws, principles and axioms.

The mathematics of the conscious psyche began the great leap forward with *The Mathematical Theory of Communication*. (Shannon, 1948) The accelerated leap forward can be noted in the expansion of ideas in the 17 years after in the introduction to the laws of information by Shannon. Evidence of the leap forward is given by Ash. (Ash, 1965) Since then the theory has resulted in the computer, the internet and an endless stream of electronic communication devices. Since Ash, there have been thousands of books and scientific papers published on the implications of the mathematical theory of communication.

For an introduction to information diffusion between the conscious psyche and the non-conscious psyche, see (Howard, 2012b)(Howard, 2012c) (Howard, 2012d)

A model of the mind function as a diffusion process

The diffusion is described: $Ln(P)=K \acute{N} F$ (approximately)

Where $Ln(P)$=natural log of P=information diffused

K=experimentally determined function

$\acute{N} F$=gradient of the potential field within the mind which is the Chi driving the information diffusion. The vitality, chi, of the mind is the gradient of potential that activates thinking. It is necessary to define the potential field of the psyche. This can be a mathematical description.

Disorder: Anesthesia, severe trauma, disease, drugs such as alcohol can stop these functions of transmitting, encoding, decoding, receiving, and processing of information.

Level 10: Ordinary Mind not taught in Buddhist methods

from which other mental state levels arise. (Defined as functions and a part of the whole world system, not as an individual independent object)

Psyche

Non-conscious phenomena such as reflexes in physical nervous system.

Mental phenomena induced by feelings.

Thought train; a succession of thoughts with a unified goal
or a definite connection between separate thoughts.

Information processing system.

Inventing meaning for the information received.

Collecting the entire volume of information being received
from body, brain and nervous system and yielding a
perception.

Encoding the meaning of information developed in the
mind for transmission to the body, brain and nervous system.

Functions of mind are defined by Heidegger in *Being and
Time*. Heidegger also defines words of vocabulary (Heidegger, 1962)

Perceiving with mind

Level 11: Mind trained in Buddhist methods yielding
unelaborated mind substrates

(The following are learned states which have become habitual
tendencies of mind)

Samadhi;

willpower focused attention to an object

willpower gathering memories

Willpower focused on not cognizing the sense data
(pratyahara)

Will power focused on non-intellectual manipulation

Will power focused on reception of information from within
body (hatha yoga)

Will power focused on reception of information from the nervous system

Will power focused on reception of information from normally
inaccessible memory (ESP)

Will power focused reception beyond habitual constraints

4 super-mundane states:

No length or space limits

No time constraints

No thinking, no cognition, no thought train, no
mental constructs (void)

No perception of self, no ego, no center

No perception of being

No presumption of knowledge from before birth, from learning, and from habitual interpretation of reality (nirvana, realization of Brahman) Consciousness of God or Brahma, Buddha Nature.

Individual consciousness expands to experience collective consciousness Individual consciousness is realized: labeled, "Atman" Buddhist Jhanas

1. Secluded, thinking and pondering, joy and rapture borne of detachment
2. Joy and rapture borne of concentration on golden flower, no thinking and no pondering, Concentration, meditation, mindfulness,
3. sole mental process is mindfulness and equanimity,
4. Concentration, meditation, absorption, neither pleasure nor pain, no emotion

Level 12: Consciousness (A channel of communicating information between senses, nervous system, brain and other entities. Many types and maybe different levels)

Awake, alert
distracted
Awareness of a sense object activates the mind.
Consciousness of touch, smell, taste, sound, appearance, memory, intellect.
Self conscious mind, contracted within the ego, introverted.
Feeling induced type of consciousness.
Identifying feeling.
Dream.
Information diffusion between physical body and mind entity and the reverse.
Information diffusion between sense organs and other parts of the physical body and the reverse.
Diffusion of sense data from non-conscious operations of nervous system.
Anesthetized.
Sick or injured
Psychosis (unable to cope with life)
Fatigued state, sleep deprived.
Kinesthetic expression of mind and thinking thru gross body movements.

Level 13: Mind function which investigates
Learning mental habits based on sense data, and existing beliefs.
Will power activated learning.
Focused thought as response to willpower.
Focused body movement as response to willpower.
Focused speech as response to willpower.
Diffusion of sense data due to willpower.
Willpower focused attention to an object.
Willpower not focused on cognizing the sense data (pratyahara).
Will power focused on non-intellectual manipulation.
Willpower focused reception of information from within body (hatha yoga).
Willpower focused reception of information from the nervous system.
Willpower focused reception of information from normally inaccessible memory.
Tool for pondering: Sustained concentration on a subject of a thought train.
Emotional contribution to pondering: Long term mood feeling
The mind levels above at the Level 1 to 13 are the thoughts, speech and action of the individual person.

Level 14: Collective mind
Some people find that their imagination and intellectual ability improves when discussing abstract ideas with a peer group. This is the collective consciousness of collective mind.

How many of the listed 15 levels of mind did the searcher experience; searching after the underlying reality experience? One can conjecture or experience that realizing more levels of mind would result in a more efficient and thorough analysis that achieved the goals of identifying and communicating with reality. Some of the levels of mind allow a more encompassing comprehension of the conditions being contemplated. Some of the levels of mind exclude the ego and the personal interests of the investigator.

Appendix E: What is Chi?

The term, "Chi" can be found in many different fields of instruction and research. It is not an exact concept but it was discussed at length before 400BC in China. (Schwartz, 1985, p. 270ff)

A particular debate in which Mencius was a participant, lasted many years. Questions were raised by the tiny intelligentsia stratum of society (shih), concerning the underlying reality. This was the Chinese era of The Warring States (481- 221BC) characterized by chaos, a state of society that has always been despised especially in China even today. Mencius and others described it as a wave moving between order and disorder. This wave motion is set into operation by one man or a small group. The shih wanted a man to arise who had the potential to initiate good.

This was the question of *"hsing"* "Where does a man seek the underlying source of moral action?

How does a man become good? This was one of the parts of the debate on human reality. Mencius wrote about the comprehensive underlying reality of human civilization in *Philosophic Anthropology*. The source of morality, *hsing*, the innate tendency toward good is latent in each person. The heart and mind, *hsin*, of man is the ultimate reality of man's nature which is different from other animals. This ultimate reality is the tendency of *hsin* to grow toward full actualization of moral ability. Chi is the source of *hsing* and *hsin*.

Chi is a vitality that binds one man to the ultimate cause of being. Chi contains courage, an emotional disposition belonging to the vitality of the psyche and body. Whereas physical courage can be used for evil, moral courage is useful for projecting a man's conviction to 10,000 others if he is righteous. Moral courage is based on righteousness and "unmoved heart," equanimity. Chi is not heart but it is related to heart. Chi is the vitality of the body-mind. It is connected to emotion, appetite, and desire.

The circulation of Chi in the body-mind is where perversions and disorders manifest. The turbulence of disorders of Chi obstructs growth

of dispositions into virtues. The dispositions are present at birth; a field of *hsing*:

a.) spontaneous
b.) intuitive
c.) unreflective
d.) uncalculating.

Within this field are the dispositions:

1.) Feeling of compassion.
2.) Feeling of shame.
3.) Feeling sense of courtesy and deference.
4.) Feeling intuitive judgment of right and wrong.

If the environment is nurturing, these disposition mature into

1.) loving-kindness, sympathy, humanity (*Jen*)
2.) virtue of righteousness in all situations (*Yi*)
3.) spirit of inner appropriate attitude (*Li*)
4.) moral knowledge (*Hsing*)

Mencius claimed that the people in power and their coterie of evil advisors cause the emperor to initiate disorder and war.

The obstruction of dispositions and the resulting lack of virtues is not the ultimate cause of evil. Deficiency of Chi is the cause of disorders.

Chi is empty and unites all things into Tao. Note that this description also agrees with the Dalai Lama and the research in physics. (Dalai Lama, 2012) and (Genz, 1999)

How does Chi come as the source of perversity and disharmony? There is no differentiation between Chi in the body, psyche or spirit. When Chi properly fills body, then all the senses and organs function in balance. But when Chi drains out, then imbalances, disorders, and massive passions occur.

If Chi is stored up., then dispersed, a man suffers deficiency. If it ascends and does not descend, it causes him to be irritable. If it descends and does not ascend, it causes him to be forgetful. If neither ascends nor descends but gathers in the middle of the body, he becomes ill. If blood Chi can be concentrated in the five organs without drifting away, then chest and stomach can be replenished, lustful desires diminish, ears and eyes are clear, and sense of hearing and sight are penetrating. This condition is called, "clarity." The five organs are then subject to heart and do not deviate. The Willpower dominates and behavior does not go astray. When the willpower dominates and behavior does not go astray, the spirit (*shen*) flourishes and the Chi does not dissipate. The willpower dominates Chi while Chi fills the body.

What are the causal relations between these categories? What causes depletion of Chi? What causes replenishment of Chi? Men of physical courage maintain their Chi by cultivating an emotional disposition to control forms of depletion, fear for example.

Connected to Chi are passions which physical courage cannot control and which overwhelm physical courage.

What causes dispersion of Chi? What causes disorder of passions which block actualization of four dispositions for achieving the source of morality, *hsing*,? Chi is affected by outer world by way of senses which are channels of communication.

Why are there great and small men? The senses are not the location of evil. They belong to our heavenly nature that we share with animals. Evil arises where Chi interacts with senses and with objects outside the body. A property of Chi and a property of sense lead to fixation or obsession or addiction or excess repulsion by things outside body. When Chi is not drained off by these fixations, obsessions, and repulsions then vital Chi is in balance. Not just evil temptations are communicated through the sense channels, but also abstract vices such as lust for power, for fame, and for honor are communicated through these sense channels.

How are senses and Chi controlled? Heart/mind (*hsin*) is the organ of thinking. Willpower within *hsin* causes thinking. Spontaneous tendency toward good is in the unreflective heart (dispositions at birth). The author Schwartz considers thought and willpower (*chih*). He thinks this is a bout (*yu-wei*) goal directed decisions; the choice between good or evil. (Schwartz, 1985 p. 273)

Mencius asserted that the heart is the organ of thinking and willpower. Heart rules Chi. Heart must focus on goodness in spite of attractions and repulsions of the world. The four dispositions at birth mature into virtues. Such a man sets the example for others.

Mencius conjectured that there must be an ideal person with "true heart" (*liang-hsin*). This person, a sage, lives in a world of harmony in his heart. He is one with the four dispositions. His senses are controlled. His Chi is nourished, in balance, and in harmony with the Cosmic Chi. He has moral courage. He leads society and the high power people out of disorder into order.

The sage can save the world of men from its tendency toward degeneration. He can free the world of men in power positions who scheme and plot, who pursue wealth and power, who have a passion for unlimited luxury, who are not governed by respect for appropriate behavior, (*li*) who act outside the law (*fa*).

This view of reality seems to be based on an intuition of the information communicated by the underlying reality through the senses and through the tendencies inherent in each human at birth. The moving vitality of Chi is postulated to be the driving cause of diffusion of many aspects of human life toward the observed wave motion between the extremes of order and disorder in society and in chiefs in government, business entities, and religions.

Appendix F: Decoding Several Levels of Reality in the Moral Mazes of Thinking, Speaking, and Acting by Chiefs in Governments, Corporations, and Religions

The recommendations and conclusions in this Appendix F are repeated from Chapter 12 and the Foreword.

Recommendations

Massive resources that have been wasted on extending the power and prestige of a tiny number of elite people it government, business, and religion for the last 3000 years. The major tool used to project this power and prestige has been weapons and war. The many tools for expanding the mentally ill egos of the elites are now threatening the existence of life on Earth. This intention to amplify the egos of a few elite people must be diverted into the intention to discover what will aid the evolution of human kind in concert with all life on Earth.

The guidance toward discovery of what aids evolution of large groups of humans is to make efforts and to spend money on the following policies.

Whatever extends all life on Earth.
Whatever brings people together under the principles of Respect, Harmony, Tranquility, and Purity.
Whatever aids the evolution of human kind, in concert with all life on Earth, toward the following.
Correct view of reality,
Correct thinking,
Correct speech,
Correct, action,
Correct mindfulness,
Correct concentration,
Correct equanimity,
Correct liberation from suffering,
Correct pleasures,

Discovery of, total reception of, decoding of, and correct use of all
communications from the underlying reality,
Discovery of the total powers of and the correct use of all the powers
of liberated minds.

I recommend the diversion of all money and human effort now
wasted in weapons, war, threats of war, threats of destruction of parts
of the Earth held by large groups of people, threats of destruction of
the means of production held by large groups of people. I recommend
this money and effort be directed toward valuable goals.

One research goal is knowledge of the communication with the
underlying reality. The intent is to discover the total reception of,
decoding of, and correct use of all communications from the underlying
reality. The sum of all the suggestions mentioned herein for locating,
intercepting, observing, decoding communications from the underlying
reality and profitably using the information are only a small part of all
possible channels and communications from the underlying reality.

The other research goal is knowledge of all the capabilities of the
human mind, especially extra sensory perception. The intent is to
discover the total powers of and the correct use of all the powers of
liberated minds.

The main objective of this book is to be the engine of directing
mental activity, funding and concentration on the power of the mind
and on consciousness based on the human body. This concentration will
be discovered to be orders of magnitude more powerful than military
approaches, than totalitarian government approaches, in allocating
human resources. Human resources must be allocated toward creating
a world suitable human survival, for positive human expression, for
human freedom, and for the survival of the Earth itself.

Conclusions about levels of reality in the moral mazes of present and past governments, corporations, and religions

Large groups of people are organized in some way to provide goods,
services, and to achieve goals. These goods and services may be provided
by governments, corporations or by organized religions. Whether the
group is a government, corporation, or religion, very similar human
tendencies distort the stated intent of the group and, sooner or later,

destroy the group. The example here is government but other types of organizations are similarly corrupt. Another type of organization, such as business corporation or religion, can be substituted for 'government' in the exposition.

The central problem is in the human mind. Humans have gained control over much of the material Earth but they have not gained control of their own mental processes. The first solution is that humans must concentrate on controlling their minds. Many spiritual masters have taught how to do this. The teachings have been neglected in the lust for material things and the mania for power. But there is a drive to power, an archetype in the human psyche that is expressed in construction and destruction.

The first objective is to channel the drive to power into construction and stop war and destruction.

Let us define the symptoms of the archetypal drive to power.

An ideal worker is absorbed in his work giving himself fully to the occupation. The reality is that some workers are absorbed in doing as little work as possible and thinking how to get more material goods and to get more power over the larger operations. This real person is ready to physically fight to release his massive energy; his aggressive need for power.

Discontented men who have gained more power may have an insatiable hunger for glory or for control over people. Many of these men may join together to amplify their power. They can be fully absorbed in destructive work but not become absorbed in constructive work. The control structures they want to create may result in a net destruction of part of the Earth. Also the mass contagion of psychic desire for war that sweeps a large group is amplified in these discontented men.

Currently, the chiefs in the world agree that the threat of war can only be countered by creating massively armed areas of Earth. However, the armed areas provoke war. This is called 'Mutually Assured Destruction' (MAD). The incomprehensible amount of MAD money and effort of millions of humans are wasted. They could be employed to create food, clothing, residences, and the businesses to supply these basic needs. Instead, billions of humans live in extreme poverty without satisfaction of these basic needs.

Love and cooperation are absolutely more desirable than war, fear, and destructive competition. Would anyone deny this?

Science, technology and universal education had promised at one time to provide the basic needs. Progress has changed everything in human life. The Pacific Ocean is now dying from man made technological creation of radiation. Humans cause parts of the Earth to be turned into desert and endless destruction of Earth. It has caused degeneration of spiritual life. It has neither changed the primitive thinking process nor the archetypal drive to power.

The archetypal drive to power causes men to ignore the needs of billions of humans, plants and animals for their individual thrill of power. The chiefs mask their greed for power behind liberation, religious enlightenment, and so on. These criminally sick power fiends have always ruled humans. They have already caused half the world to be destroyed not just by war but by mining, fishing, damming rivers, destroying forests, and so on.

Part of the problem is that the overwhelming majority of humans have not put a stop to this. The lack of action by those who could have acted, the silence of the voices of justice when it mattered most, has made it possible for the evil chiefs to triumph. The people must find means to stop the insane and the criminal people from being put in control. Democracy is the easiest avenue for the evil ones to get control. An appropriate method must be invented and used to select chiefs.

In addition to testing children for reading and arithmetic, they must be tested for morality, interconnection of all life, and stable sanity. Those who pass can then be educated in peace, limiting an organization to its stated objectives, cooperation, negotiation, and administration.

We need new thinking. The minds of the chiefs must be screened to prevent the domination of war and the drive to power as their motivating thoughts. For example, the Department of Defense or the War Office must be transmuted into the Department of Peace and Cooperation. (Hutschnedker, 1974)

The problems listed below suggest that a new mental construction of how large groups operate is needed. New principles of operation are briefly offered at the end.

The central purpose of all government

One central problem of government is to enable the material basis of all people in a large group. The problem facing a large group is producing the material basis of food, clothing and shelter. In addition, there are services that are efficiently delivered by specialized places or by specially trained people. Education is a service that is allocated to local governments not to the Federal government. Education has always been considered to be the job for government although there are private schools and colleges.

Government is composed of people. People commonly have sinful thinking, speech, actions and mental illness. People attracted to holding positions in elite government are damned with more extreme mental illness. They tend to have distorted cravings for wealth, for the power to change the structure of large groups of people. They may feel, for example, that the exercise of power will satisfy their cravings, power is the cure for their mental defects. Their delusion is that the cure to their mental defects lies in their clinging to controlling the thoughts, speech and actions of the large group. The exercise of power yields a mania to exercise more power: power addiction is not curable.

Government and other large controlling authorities are usually the source of destruction of society

Sooner or later, the ruling elite has always been the enemy of the evolution of human kind. Governments have always ended because the great majority of human kind in a particular group as a whole removed the governments.

Contrast the Protestant Christian work ethic with the bureaucratic government lack of ethics

There is a myth about the underlying reality as interpreted by the individualistic Christian business owners during the golden age of America. The interpretation of the underlying reality from about 1600 to 1960 was the American ideal. The ideal man owned his business which might include a farm. The ideal was not to be an employee. The interpretation of the communication from God, the underlying reality, was the work ethic. The Ethic was a mental construction of a self-confident, energetic, ambitious, and independent man who owned his land and everything on it. He was tied to the land and his family accepting the stewardship of all of it. The integrity of the handshake as an inviolable contract was crucial to business relationships. Hard work and honesty led to success and to the blessings of God. If someone was miserable and poor, it was a sign that God had forsaken him because he was not working hard. One felt strong guilt if one did not work hard, if one violated ones handshake, if one did not support the ideals of the Christian government, and was not rewarded by God with salvation in Heaven. One learned the rules of the underlying reality, God, from the Bible in Sunday Church. The rules were built into the one's mental reactions to the world.

Government bureaucracy undercuts the Protestant work ethic. The bureaucrat is not motivated by the promise of God to be rewarded for hard work, and for honesty. Property ownership is not part of the obligation. The bureaucrat is not dependent on God, nor does he seek the rewards from God for hard work, nor does he feel guilty if he does not work hard or if he is dishonest. There is no connection between work and salvation by God. He is dependent on the whims of his boss and the market buying his services. The rules of the bureau take the place of the Word of God in the Bible. This is a different interpretation of the underlying reality which was often not connected to God. His rules were dependent on the situation and on his falling before the temptation to acquire personal gain through twisting the administration of power for his private benefit.

The contrast is between a man who perceives God behind the government rules and a man who perceives that the bureaucracy of government is capriciously making up the rules which can be made to enrich the man.

There are endless interpretations of the perfect or the practical rules of government. In spite of all the variations of mental inventions of government, one can conclude from reviewing history that government has rarely been adequate to yield satisfaction of the material basis: the necessary goods and services required for the orderly continuation of a large group. To the contrary, the people who make up the governments have always been addicted to their own personal material things, power, prestige, and so on.

Due to the archetypes, tendencies and mental disorders built into the human mind, all governments fail: some in less than a year. China had two governments, the Song dynasty (960-1279AD) and the Ming dynasty (1368-1644AD) that came closest to the idea of providing the necessary services and goods. Each of these lasted about 300 years. The Ideal of a republic of America served well from about 1800 to 2008 when the dictator Obama conquered the Federal government without opposition. Rome had excellent government in combination with religion and business interests from about 500BC until it began changing into an empire in 60 BC.

Another conclusion is that, currently, most governments in the world are preoccupied with war and with projecting the power, possessions, and prestige of the ruling elite. There is a major expenditure by governments and private industry for war including but not limited to money, material, research into technology, and educating humans to perpetuate war. The entire Earth is being expended in the employment of various types of thinking, speech and actions to extend the power, possessions and prestige of the elite people in government, the elite people in business, and the elite people in religions. This diverts the resources from meaningful evolutions that would benefit human kind and that would preserve the Earth as a whole.

Government must provide more than basic goods and basic services

Another central problem of government is how to nurture the human archetype that emerges from a large group. This means encouraging positive or harmonious group behavior and retarding destructive or discordant group behavior. Creative and destructive group behavior will be addressed below in the light of ancient history.

In addition to material basis, education is also required. The many governments have perverted the education process by refusing to teach what is necessary and teaching propaganda and lies.

The spiritual basis of a person must be woven into the general education. The major objective of education is to purify the mind and to provide an avenue toward avoiding suffering and toward enjoying pleasure. All human action and speech originates in the mind.

Educating how to avoid mental hindrances and how to cultivate mental enlightenment factors are the primary objectives of education.

The Ancient Problem of nurturing the human archetype that emerges from a large group: encouraging positive or harmonious group behavior and retarding destructive or discordant group behavior.

The large group of Asians in what is now China contained special intellectuals who tried to invent a perfect government. The invention of the ideal behavior of the large group, the state, was invented by the Chinese beginning about the time of Confucius. Varieties of implementation of a government that provides the material basis has been tried since then.

A government that provides these benefits had been implemented in China. It only lasted perhaps 300 years in China during the Song Dynasty (960-1279 AD) and again about 300 years in the Ming Dynasty (1368-1644 AD). The American republic lasted from about 1800 to 2008. The Roman Republic lasted from about 500 BC until 60 BC.

The problem of providing these material basics and also education for large groups was recognized by Confucius (551-479 BC). Confucius grew up in the province of Lu which was governed by three brothers who had ejected the previous ruler but kept him alive as a proof of their authority. The province was overshadowed by the intrigues and jealousy of the three brothers, and also by heavy taxation to support their extravagantly luxurious lives. Their executive ministers originated deceit, violence and massive corruption.

Confucius invented an ideal society that has been the model upon which all subsequent societies added elements

Confucius easily identified the defects of this government. He framed principles of government

1. All government efforts must be for the objective of providing food, clothing, and shelter to all the people, not just the rich, the elite, or the powerful, in a peaceful environment free from war.

2. Only the most virtuous men must be selected to lead. They must be selected from the entire population, not just the rich, the elite, and the powerful.

3. Virtuous men must be created by widespread education in the skill of government for all people.

These principles would theoretically yield a prosperous province.

Confucius hypothesized that humans are born with the potential of growing into virtuous beings: considerate, honest, reasonable, and benevolent.

Based on this hypothesis, the corollary is that they would cheerfully bear misfortune because they would also share good fortune when it came. Also people would support a government based on his three principles. In such a country, there would be an end to the need to conquer, for war, and for injustice.

Social advancement needs cooperative work. Cooperation not fighting is needed for any given person to rise above his ignorant state. It is needed for the whole society to advance beyond primitive group behavior. If people were dealt with in a just way, they would give their whole effort to help the group as a whole. Such a solid society run by men living by the three principles would realize what Confucius imagined as the highest level.

All this development of the group would depend on leaders who would live by the three principles listed above.

A major problem was to invent the process of selecting the leaders who would pursue the three principles. Confucius was convinced of the nebulous concept of 'the virtuous man.' The virtuous man was never defined adequately. He was postulated to have the inherent ability to recognize other virtuous men. A virtuous man was postulated to exist in the position of duke or king. He would be guided by his virtue to select other virtuous men who would pursue the three principles. These virtuous men would act in his name, to govern the various functions of government, and to govern the provinces far from the king.

No detailed rules or laws were necessary in selecting virtuous men because the virtuous would rely on the rules and not on their sense of responsibility to think for themselves and their tendency to be true to their virtue.

Compare the ideal of Confucius with later ideal societies

In Greece, Plato invented a similar government about 40 years after Confucius died. It can be found in Plato's book, *The Republic*. Other Europeans invented similar countries of the mind. One was Leviathan; another was Utopia. The American founders invented the Constitution of the United States of American based on some of the concepts of Confucius. (Levin, (2012)

Let us contrast the ideal of the Confucius government with the ideal of the American democratic system. The American system of selecting a virtuous man depended for a while on all property owners voting. This is based on the hope that the majority of property owners would recognize the virtuous man. This was a vain hope which never materialized. In addition, later the Americans let anyone vote whether he is man or woman, whether he is an American citizen or not, and whether he is educated or not, and whether he has a Higher Power to guide him. Thus, the voters were totally ignorant of whom they were voting for. They elected many criminals who dismantled American foundations.

Currently, almost anyone can run for elected office. The election process is recognized by criminals and people with mental disorders as a relatively easy way to gain enormous power and money. Elections attract morally defective people. Hutschnecker, a psychiatrist with a medical doctor degree, has been studying the various types of mental disorders afflicting elected people. (Hutschnecker, 1974) American democratic system of elections has produced Federal, State and city governments that are gangs of criminals.

Chinese pondered how to create a government that was not liable to be corrupted by human archetypes of selfishness and extreme abuse of power

In the centuries after Confucius, many brilliant Chinese realized it was not possible to start a government by finding a virtuous king. For example, Hsün-tzu codified laws that recognized the reality that officials representing the emperor were the worst violators of the Confucius' principles of governance. Later proposals recognized the reality that the emperor was usually the source of disorder in society and the cause natural disasters.

They formulated rules to curb the selfishness and incompetence of kings and emperors. In our contemporary America, we would call this 'belling the cat.' Who would present the rules to the king and who would enforce them? No one was brave enough or willing to die at the hands of the king's guards.

To this day, no one has ever been able to convince the emperor, the rich, or the holders of the top power in the Chinese Communist Party to obey laws. The denial of apparent reality has continued for several thousand years.

Benjamin Schwartz described a mental archetype: the manic pursuit of the perfect society.

Schwartz observed a type of thinking by certain type of people. The problem with the thinking is the refusal to accept "apparent reality." Due to this refusal, a group may obstruct scientific adaptation of a larger group. A group who refuses to accept apparent reality may also invent another reality that does not yet exist, such as a pure communist society. This type of idealist or extremely ambitious person is pervasive in many large groups such as whole nations. Apparent reality demands that one or more humans are bound by culture, by their short life in history, by social position within strict social levels, by the limitations of the human mind, and by other forces that retard adaptation. (Schwartz, B. 1985, p. 6-10)

But most humans believe their behavior and their expectations of a future world, which are actually based on their individual beliefs or their group convictions, are coming true or will be true if they force it to come true. There is a tendency by some people to ignore apparent reality. They are a group of delusional people pretending that in all cases that their mentally invented world of how life ought to be has replaced apparent reality. This can be seen in communists, socialists, the American Democratic political party, and other idealistic groups. The riddle is "Will their idealistic hopes for a world of their dream ever exist?" How does one answer this riddle?

There exists in most large groups two distinct convictions. This is another example of mass delusion. About half a society believes in the ability to create goods, services, wealth even though the people are inherently sinful. An acceptable explanation of this creation is the invisible hand of the capital market.

The other half of society believes that a capital market leads to selfish and criminal acts. They believe that because people are sinful there must be a government that controls goods, services, and wealth in spite of the fact that governments are composed of sinful people. And in spite of the historical fact that only three governments, two Chinese, the Song and the Ming, and the American from 1800 until about 2010, have ever enabled prosperity or even minimal sustenance when given the power to do anything they want.

Of the many people who deny apparent reality, these people claim that their "truth" is reality and in addition, their unsupported claims of truth" are reality. They also claim their "truths" do not contradict apparent reality. They claim their truths are not based on wishful thinking, nor based on their desires for their own individual selfish future benefit, nor based on their refusing to allow the government to control their personal goods and interests. The result is always the same. The government, full of sinful people, directs a vast amount of goods and services to its employees and its managers. The result is that the great mass of people is reduced to poverty or worse while the government people become rich.

Examples of failures of large groups or even entire cities to decode the communications of the underlying reality

Feng Shui is a Chinese invention which intends to intercept the underlying reality and to make building or cities that are in harmony with the natural laws of reality. This is an over-simplified description. The opposite of Feng Shui is the denial of underlying reality. In denying natural laws of economics, metallurgy, and farming, discord is engendered in society. The denial by a large group is a type of noise introduced into the facts and sense data.

Thousands of years of war have left the record of a human group phenomenon which cannot be denied. The evidence is that humans cause endless war and that making war is a human mental archetype

War is communicated to humankind as history, as autobiography, as travel writing, as narratives by those who were there. Each of these classes of writing history is quite different. Much of war is presented in delusions conjured up by those who were not there in the mud with a missing leg having just killed another man. They glorify the murder of about 300 million young men in the last 100 years. They deny that the future leaders of humankind have turned into ground meat. They ignore the approximately 500 million men who were mentally destroyed in war and will never enter what others call reality. The grandest and most noble lie was, "This is the war to end all war." (Hynes, 1997) (Fyfe, A. J. 1988). *Understanding the First World War: illusions and realities.*

Conclusions about squandering scientific effort and wasting the required resources and money on war

An enormous and extremely expensive effort has been carried out for over 100 years by physicists to decode the expression of the underlying reality. The experiments and observations were searched for the messages about the construction of the world out of energy, particles, and waves. The summary above demonstrates some of the progress in investigating sub atomic particles. All have mass. Light has mass which is so small that it is treated as zero. An interesting discovery is that particles, time, and space can be created out of the fluctuations in the void. This was noted by a Taoist master, Lao Tzu, thousands of years ago. This was also embodied in creation myths all over the world.

The efforts of perhaps a million of the smartest humans have been expended in the last 100 years on conjecturing on the origination of matter and the description of material sub-atomic particles. This profligate waste of creativity has been concerned with finding and describing the nature and mathematics of the final substrate of mass and the entire set of properties of mass. The latest decision of the final substrate is called the Higgs Boson discussed in the main part of this book.

The research resulted in the worst possible weapon of war, the atomic bomb. A nuclear electric power plant in Japan is now threatening to destroy the Pacific Ocean with radioactivity.

Research into the vast power of the human mind has been neglected in favor of research into destructive weapons of war

Scientific research intends to denigrate questions on the origination of the human mind and consciousness. There has been extensive research into the material brain and nervous system. However, almost all scientific endeavors ignore the description of the nature and mathematics of the non-material substrate of mind and consciousness.

An enormous amount of intelligent research, millions of careers, as well as billions of dollars, have been expended to study the chemical basis of life. Other massive scientific research programs continue. Many sciences intentionally reject the description of immaterial mind and consciousness. Some effort has been made to research the electromagnetic fields generated by the human body. In addition to this effort, consider the worth of research into either the psychic field or the consciousness field.

It is necessary and sufficient for humans to make a massive relocation of human resources to concentrate, to meditate, and to be absorbed in mind, consciousness and vitality. In fact this allocation of resources is necessary to prevent the disappearance of human kind.

The payoff due to re-allocation of war resources into human mental development especially psychic development

The question on the forefront of the study of brain-consciousness-mind is, "Can one formulate psychic particles, waves, or another form in analogy to the subatomic particles, waves, or another form?" Another question is, "What is the potential benefit from the intense research into the human mind, the psychic powers of the mind?"

Another question is, "What would happen if the majority of people on the Earth began to obey the teachings of the spiritual masters? What kind of interconnected life on Earth would exist if the majority practiced Buddhist pursuit of purification, concentration, meditation, and absorption?"

A society that denies apparent reality, pursues idealistic system such as communism, also simultaneously denies the failure of the idealistic system. These societies fail to provide the basic necessities expected from government

The denial by a large group is a type of noise introduced into the facts and sense data. This process of denying the apparent reality and also denying the failure of the newly imposed system is not unusual.

History recorded the hysterical invention of many perfect systems of rules for large groups. The idealistic system may be that half the society pays for the other half in a welfare system as practiced in The USA. These systems promised that all human needs would be provided for. Often, they promised that less work and no exchange of money would be necessary because of the vast surplus of goods and services that the system would provide. Many Chinese believed the promises of a communist future perfect society. They became excited by the conviction that they have formulated the exact system that will provide the perfect government and thus the perfect civilization

The presentation of each system might be communicated by large groups of screaming and crying people in a religious frenzy. Large groups would dedicate their lives to implement each system. This has happened for Communists, Socialists, Democratic Party in USA, German NAZI party, Muslims, and the Catholic Church.

This system of promises of a perfect future was observed by Thomas Mann in the Society of Jesus, the Jesuit subset of the Catholic Church. It intends to teach the entire human world of the benefits of totalitarian government lead by the Jesuits. (Mann, 1964)

The evidence is that a government based on idealistic mental conjecture, such as socialism, communism or the American nation based on welfare, and eternal warfare, always fails to achieve the idealistic goals.

The Golden Age of China

There is persistent myth invented by many societies all over the world: the Long lost Golden Age in China, this Golden Age was imagined to be associated with the Yellow Emperor about 2698 BC. Another person associated with this age was the son of the Duke of Chou about 1027 BC. The son hypothesized the concept of the Mandate of Heaven. Heaven puts the emperor, duke or baron in place. When such a ruler loses his virtue, Heaven will cause revolt or another cause such as an earthquake or typhoon to remove him. The son hypothesized that a ruler must employ principles of humanity and justice when dealing with the great mass of peasants and even those who are rich, elite or powerful.

Chinese history is a process of change. Government must change to suit later thinking. The Chinese, called Legalists, recognized that the Confucius form of government was not possible. They invented a system of laws: punishment and reward with emphasis on punishment. Crimes were not defined except that they were whatever the Emperor said or whatever the local chief said.

Contrary to Confucius, the assumption was that people are born evil. They must be controlled. A tiny number would do good things.

The laws freed the Emperor from having to be wise and from judging many cases. They freed him from having to be virtuous because the laws would prescribe the punishment. The Emperor had the duty of being an example of virtue for the people to follow. If the Emperor were not virtuous, the nobility and the common people could use the laws as an example to obey. The law required that the nobles reduce their arrogance and their crimes against the common people.

The Mandate of Heaven would allow the dynasty to exist if the Emperors were virtuous and competent. Otherwise, the typhoon would blow, the earth would quake, the rivers would flood and rebellion would arise.

For more, see (Kruger, 2003, pp 64-85) *All Under Heaven: A Complete History of China.*

Government principles based on self deception and on convictions without basis in reality

Confucius denied apparent reality. He transmitted the way of the ancient gentlemen from hundreds of years before him whom he believed lived by the following code.

Chih, uprightness and integrity.

Yi, righteousness, doing the right thing.

Chung, loyalty and consideration of the feelings and needs of other people.

Chu, do not do onto others what you do not want done to yourself.
Li, deference to superior in rank, son to father, official to emperor,
wife to husband, younger brother to elder brother, woman to man,
correct etiquette.
Jen, loving-kindness, empathy, inner intuition leading to sympathy
for other living beings.

This reality was based on wishful thinking, based on unproven
conviction; the conviction that long in the past the kings were virtuous.

Later, Mencius and others codified laws of reward and punishment
that recognized the reality that people often violated these unproven
convictions. This was closer to apparent reality. Later, Hsün-tzu codified
laws that recognized the reality that officials representing the emperor
were the worst violators. Later proposals recognized the reality that the
emperor was usually the source of disorder in society and the cause
natural disasters. To this day, no one could convince the emperor or the
holders of the top power in the Chinese Communist Party to obey laws.
The denial of apparent reality has continued for several thousand years.

There are recognitions of levels of reality from long before Confucius
until today. Some of the levels of reality are examples of massive
conviction of imaginary wishful thinking or delusional reality.

Many people deny apparent reality. These people claim that their
"truth" is reality and in addition, their "claims of truth" are reality.
They also claim their "truths" do not contradict apparent reality such as
economics and agriculture, steel making and farming. They claim their
truths are not based on wishful thinking, nor based on their desires
for their own individual future benefit, nor based on clinging to their
personal goods and interests. (Schwartz, 1985, p. 6) In fact, the new
world order is indeed based on clinging, individual benefit, and the
interests of each individual man in spite of the damage to society as a
whole.

The concept of human archetypes

Carl Jung and others observed repetitions of certain behavior in
humans. They hypothesized that there is an inherent behavior woven
into the human brain-consciousness-mind that is expressed in modes
of behavior in individuals and in groups.

One could explain that the source of human group behavior lies within the individual body, consciousness, and mind which result in group behavior. One could label the universal evolution of human group behavior as 'archetypes.'

All these ideal systems ignored human archetypes that defeated their implementation into reality. The Confucius system, and the other ones, ignored the human archetypes toward war, toward the unjust selfish acts of any powerful person, the corrupting influence of power, and the absolute corruption of absolute power.

The Golden Age in America

The Christian Bible imagines a golden age called the Garden of Eden. Americans imagine the golden age in the congress of the Founding Fathers 1775 to 1787 when the Constitution was framed. Americans fondly remember the American Golden Age as 1775 to about 1945 after America was promoted into world wide empire. The human archetype, the drive for power, tarnished the Golden Age. Symptoms of the failure of America are the election to president of the illegal immigrant Obama and the widely accepted principles of Karl Marx.

Except for the original Americans and the original Australians, society has ever succeeded in implementing sustainable living conditions much less Confucius' ideal country government for more than about 300 years.

One could oversimplify that the state is composed of the governing elite and the common people. However, one must admit that another human archetype is the evolution of many levels of the hierarchy. For about 400 years, from about 500 BC to100 BC, the Chinese continued to attempt to make group behavior harmonious and productive of adequate food and other products in a peaceful environment not war.

Finally in the Sung dynasty (960 -1279 AD) and later in the Ming dynasty (1368- up to 1644 AD, the Chinese developed two of the finest governments in the history of human kind.

Other excellent government systems were the American republic from about 1800 to 2008. The Roman Republic lasted from about roughly 500 BC until about 60 BC.

The Europeans have never invented a successful group behavior that lasted more than a few years.

Another invention of a society similar to that proposed by Confucius was produced by Karl Marx in about 1840. It was adopted by the Chinese about 1958 after China had been in revolution, civil war and Japanese conquest since 1911. The farmer revolutionaries needed a philosophical foundation for starting a new order of government. It was implemented neither in accord with Confucius nor with Marx. It was based on Stalin's perversion of the Marx concept. The actual Marx concepts did not reach China until after Mao died. The new Chinese ideal government was learned in Moscow by visiting Chinese trainees.

[A note on labels used as noise to prevent accurate decoding of signals: Americans, Europeans and even Mao labeled himself, 'Communist.' Salisbury thoroughly examined the revolution and the People's Republic in China from about 1921 to 1968. He showed that Mao was a farmer revolutionary as were most of the warlords rebelling from 1921 until 1949. The label, 'communist' was a useful noise that Mao used to kill his enemies and to cling to power. (Salisbury, 1969), Chapter V. Is Mao a Communist? p. 67 ff.)]

Mao repeated the pattern of trying to conquer the country and gain personal power under the noise of transforming it into an ideal condition has been repeated in human history. The delusion of the ideal human government is described briefly above by Benjamin Schwartz.

Exception: the ancient original peoples in the Americas and Australia were able to invent successful large group functions that lasted for 1,000 to 50,000 years.

The original Australians lived in a sustainable society that lasted about 10,000 to 50,000 years. This was destroyed by the destructive Europeans who systematically murdered them.

Some of the original people who lived in North America lived in sustainable systems that lasted about 1,000 to 10,000 years. The systems were destroyed by the Europeans who systematically murdered them. Soustelle provided a brief account of the destruction of the Aztecs in *Daily Life of the Aztecs: On the Eve of the Spanish Conquest*. (Soustelle, 1970) There are numerous books giving the details of the genocide of the original people in North, Central, South America and Australia.

Constructive and destructive archetypes in
some native American cultures

The underlying reality of a human archetype drives a large group of humans to become a living, self perpetuating interdependent organism, the village, tribe, and city. Part of the archetype includes constructive group behavior: common decency, family traditions, moral and legal codes that are responsible for the development of a satisfactory and enduring society.

There is another part of the underlying human archetype that drives the same group toward willing destructive group behavior: degradation, degeneration, immorality and especially gambling. These two opposing tendencies of the archetype that yields group behavior tend to destroy the village, city or culture.

These two group behaviors were recorded by the American original peoples. These two extremes may occur in all large groups of people.

One of the probable causes of the disappearance of the village or city organism is an upper limit to the number of people that compose the city, 'critical mass.' The recorded oral history of some American original people emphasize that when a group of humans reaches a certain size, 'critical mass' then the archetype of selfish behavior becomes a large scale kind of counter productive behavior. Then the destructive archetype is preponderant over the constructive tendency. The archeological record agrees with this history.

Page reported that the ancient American original peoples recognized the archetype that destroyed their highly developed cities. The archetype was passed on for over 1000 years in the oral history. It was verified by work of the physical anthropologists exploring the remains of their cities. (Page, 2003)

In the anthropology of the original Americans, it is obvious that many highly developed cities were abandoned. Usually, there is no indication of war or lack of water or other resources. Inquiring of the remaining people near the abandoned city, the anthropologist turned up the following analysis explained by the oldest living original people in North America. (Jake 2003, p. 65-88 Chapter 3 High Society)

Throughout the Southwest of North America, cultures had fallen apart. People were leaving ancient sites, and the associated old ways. They amalgamated into new tribal people. What were the causes?

There was a cyclic climate change that reduced the amount of water, caused unbearable high temperatures, and increased the evaporation of groundwater. This caused a disruption of the rhythm of living and resulted in emigration. Hopi clans can even now remember prophecies telling them to keep going until they reached the final place after centuries of this pattern of emigration.

The human mob was driven by the destructive grasping, craving, clinging, gambling, and the addictions of pleasure. These human archetypes have always been at the root of disintegration of cities.

The ancient original people of the Indian continent also passed down this legend of the destructive archetype. The devil god, Shankhasura, hid the essence of wisdom and the blueprint of harmony in the ocean mud. The result was the fading away of human discrimination and knowledge of correct conduct. Gradually greed, selfishness, fear, sexual desire, and self-gratification were the human goals. Heaven, Earth and Man moved far apart. To appease the massive and insatiable lust transferred to humans from Shankhasura, humans plundered the Earth, wasted the forests, polluting the oceans, making deserts of the land, filling the atmosphere with poison smoke. The human population increased exponentially. but human health was poor and vitality was low. A few people had almost all the desirable material things while virtually all humans and animals lived short lives in extreme poverty, suffering, sickness, and pain.

The living residents of Heaven hid in caves. Shankhasura caused earthquakes, floods, devastating fires, drought, and epidemics. The enlightened human sages begged Vishnu to intervene. Vishnu told them to retrieve the wisdom and knowledge from the mud and meet him at the intersection of the Ganges and Yamuna Rivers. Vishnu framed the continuous ceremony of interconnectedness which upholds the order of the universe, the matrix of the cosmos, and nourishes the web of life. All humans must trade in their personal desires, harness their minds, and control their sense desires in return for the common good. The formula is that all humans must give their individual everything in return for harmony among all people. The Holy Vedas tell how to do this. The Christians will claim the Bible tells how to do this. Other religions have other claims. The many claims lead back to conflict and the lead back to religious chiefs taking everything for themselves.

Each year for about 5000 years millions of Indians go to the intersection of the Ganges and the Yamuna rivers to wash off their sins.

There are phases of cooperation, protection of the Earth, prosperity, peace, and harmony through India. The next phase is another fall into the worship of Shankhasura.

This pattern repeats just as the original plains peoples described the cycle of human group behavior. in North America.

In contemporary America and Europe, these same destructive behaviors are observed. For example, The American national profit (GNP), consists of about 20 to 30% of financial profits. Most of financial industry activity is gambling on a massive scale, neither producing goods nor services, nor adding to the GNP. Financial profits are not construction or manufacturing or professional services like education or medicine. They are manipulation of money at risk: gambling.

The community of original Americans ignored the collective memory of pain and suffering which caused the last disintegration of their previous abandoned city. They ignored the need to expend more effort on the satisfactions such as food, clothing, shelter, education, family refuge, spiritual expression, and community construction projects. Within the collective orally transmitted memory, from 1300 AD onward, was stored the following group pattern. From human nature, emerged the following archetype.

The effort towards satisfying physical comfort produced an excellent community foundation. Then trouble would always begin. Men, more than women, would pay less attention to the agricultural fields, the source of most of the food. They would assume the necessities were available without their efforts. Instead, they would gamble. Women, who were the dominating sex, would play sexual games with other women's husbands. The common people and even the contemptible ones would lose their humble deference and would become increasingly arrogant, vain, self important, even proud. This is now the degenerated condition of North America, Mexico, South America, Europe and China.

It became clear to a few, who recognized the principles of building wealth, that there was a need to recover the old ways of maintaining the order of community, of purifying the decay, of calling on the gods for help. These few called out to the remaining upright people to return to the old ways that had yielded the foundations upon which their prosperity flourished. They asked the gods to guide them toward the old successful ways. The gods replied by flooding the entire city and the surrounding untended fields. Most of the people were drowned. The few who survived told of the decline of the old ways which led to the destruction of the city. The Christian Bible records a similar event in the book of *Noah*. Noah built his ark to save animals from the flood which was sent by God to punish the sinful people.

In China, floods were blamed on the sinfulness of the Emperor.

This collective memory recalled this destructive archetype many times in several different groups. The Christian Bible records the same archetype in *Jonah*. Jonah was sent by God to order the sinful people to return to worshipping God. (Anonymous, 1910d)

Note that this pattern was observed by anthropologists all over North, Central, and South America, in the Mayas, the Aztecs, and the Incas. After an excellent functioning city or regional community was founded, trouble started.

Strangely, the North American plains tribes did not appear to degenerate after many centuries of existence.

This constructive and destructive pattern was also observed in the Chinese Empire, Middle East, British Empire, Roman Empire, and Ottoman Empire. Europe has borne this pattern repeatedly for 1000 years culminating in the World War I and World War II. Now the American Empire is disintegrating for the same reasons.

Constructive and destructive archetypes are expressing now in contemporary American and European cultures

The trouble began in America about 100 years ago. That is when President Woodrow Wilson and millions of Americans craved to be in the European war; World War I in which there was no American risk. That was the beginning of endless American wars against other countries. Financing endless war raised the vast public debt that is the ultimate weapon which is destroying America. Kennedy described this self destructive pattern of empires in *The Rise and Fall of the Great Powers*. (Kennedy, 1987)

Foreign wars destroyed the European nations. The Roman Empire was destroyed by the debt and other factors resulting from distant wars. Other causes accelerated the destruction caused by war and massive debt: religious war, witchcraft, etc. A major feature in destruction of nations was the elaboration of things or excessive hierarchy. Now, this same cycle can be observed in the American Empire.

In an empire, the religions often elaborated into cathedrals and excessive hierarchy.

Science was elaborated into a religion which expressed itself in the Large Hadron Collider in Switzerland and France. The collider is about 23 square kilometers of underground cathedral and associated surface facilities. Excessive taxation was needed to pay for the endless hierarchy of officials.

The American Republic began its transition into the American Empire when it entered World War I with the intention of making the world safe for democracy. The rush into empire accelerated during World War II. The American Federal government changed away from the Constitutional Republic as liberty was slowly smothered. The archetype of destruction of society has also accelerated.

Ownership of technological possessions and prestige items has replaced individual spiritual and religious development

The degeneration of America away from industry and agriculture can be observed in the following symptoms. Drinking wine and other alcoholic addictions has elaborated into extremely expensive varieties of drinks. The game is to profess one's expertise on alcoholic beverages. This is a mask behind which the experts get drunk.

The need to feel massive pride has elaborated into ownership of wine, art, baseball cards, books, and pleasure craft such as elaborate boats and jet planes. These are held as the masks of pride. They cost $1 million to $200 million and are claimed to be reasonable investments.

The excessive elaboration of the latest technical gadget is demonstrated by high definition television, smart phones, ipod, laptop computers, kitchen appliances, trash crushers, massive refrigerators, and food processors. Many are addicted to the smart phone and the TV. The lies and propaganda delivered on the smart-phones and TV are believed exactly the way the dogma of the Christian Church was believed.

Cars have been elaborated into symbols of status, specialized sport equipment, massive shelters, and pleasure palaces. Some cars are carried in the private jet plane with its owner.

Political election campaigns are elaborated to cost US$ millions to US$ billions.

Modest weddings are elaborated to cost more than $30,000 and the bride's dress costs from US$3,000 to US$100,000.

Women's clothing is elaborated into status symbols. Purses can cost $50,000. Shoes can cost $10,000. Women pay a premium of $200 more if a pair of humble denim jeans has been overly bleached and holes cut into their knees. They pay $2000 to $10,000 if the denim was owned by a celebrity or important person.

There is a strange inversion. Entertainers such as baseball players are paid about 10 to 100 times as much as responsible professionals such as medical doctors, state governors, and rocket scientists.

These elaborations are symptoms of a collapsing society. The emphasis on what is necessary for a healthy body, a spiritually attuned mind, the refuge into spirit and religion is almost gone. One can witness the archetype of the destructive cycle in at least 50 major changes away from America the republic as founded; away from America as defined in the Constitution, into America the Empire.

Symptoms of decay are the purges, preoccupations with trivial issues and purposefully ignoring the unfolding disaster

Paul defined 50 major problems which have arisen due to the decay of the American Republic into the American Empire. (Paul, 2011) In addition, the following symptoms are a few of the many Empires that ended in the disappearance of a thriving civilization such as Russia, China and now the United States of America.

Since about 1955, America has been in the midst of a purge over a trivial issue. It is a vicious racial prejudice against mostly white people. The myth is that white people have black slaves. This has not been true for 150 years. White people are being purged from positions of power and publicly humiliated.

The national debate on racial equality has become a tool to ruin the lives of white people. Blacks have been elevated into the top social ranks regardless of their individual sub-human and criminal behavior. Whites are generally sneered at by the colored races although the non-blacks are not allowed to sneer at the blacks. Race is a mental construction and impossible to define. There are endless gradations and mixtures of races.

In spite of the many levels of unequal race categories, there is the general propaganda of American equality of all people. This is impossible on its face. People have too many variations to be equal.

There is a substantive conflict between the original American people and all others. This remains as the original American race trouble. The genocide practiced by European descendents on the original Americans is collectively remembered by the original cultures. The focus on this artificial race problem diverts away from the significant problems and their solutions which are destroying the country.

The Chinese under Mao used name calling, "rightist, "revisonist," "capitalist roader," and "landlord" as excuses to murder tens of millions of Chinese who threatened the supreme power of Mao. The name calling and murders resulted in China's suffering decline in all pursuits of life except the power of the Chinese Communist Party including Mao and the power of the military. The decline stopped when Mao died and the name calling stopped.

Stalin used the same name calling to murder people who threatened his power and the power of the military. The USSR declined in all other parts of life except the Russian Communist Party, the armed forces, and the pervasive secret security forces. The decline continued for 40 more years after Stalin died.

Now America is using name calling. It exhibits the symptoms of decay. It is suffering massive decline in production of wealth, in trust of government, and even in military power. But the secret security forces are multiplying rapidly.

The fascination with technological products and material possessions diverts efforts away from solving the substantial problems of America

America's self destruction is masked by empty elaborations of all types. America and its immigrants crave the status of fancy things or exclusive events. Does this make them satisfied, happy, and joyful?

No! The faces of people driving cars are determined, hating, proud, jealous, snotty, and impatient. They lack the grace of the refined poor. They lack the serenity of the Buddhist monk. They lack the wealth of harmony.

In the first centuries of European colonization of the Americas, friendship was largely based on cooperative work. It was called the Protestant work ethic. Americans no longer frame friends in the concept of cooperative work. They do not share the joy of others in sympathy. They seldom initiate loving kindness in a difficult situation.

The conversation of negotiations has been replaced by analogy: the screaming of the spoiled brat sending a lawyer to settle in court.

America was once based on capitalist principles. Currently, who disciplines their children in the principles of building wealth? Very few, perhaps one in a hundred. How many people plan for 20 years ahead? About one in a thousand. For one year ahead? Perhaps one in ten.

Who remembers a need to recover the old ways of maintaining the order of community, of purifying the decay, of calling on the gods for help? There are few of the remaining upright people. The degeneration remembered by the American original peoples is happening in America today. How many call for a return to the old capitalism and the Protestant work ethic, ways that had yielded the foundations upon which American prosperity flourished. Who asks the gods to guide America toward the old successful ways? Who calls on the gods to restore the old ways? Out of a hundred Americans, how many pray each day or attend to the congregation in a church or to the Sangha in a Buddhist temple?

Now is the time for an appropriate organization of government, corporate system, religious system or a new system to provide all people with basic goods and services

This year, 2014, there are many spaces on the Earth that used to be countries but have neither one recognized government nor a representative entity such as a dominating, church. Wealthy drug barons run Columbia and Mexico. There are countless civil wars. These are symptoms that there is no practical system to provide necessary goods and services. Such systems used to be provided by governments, churches and private companies which are now breaking down.

America is a dictatorship with endless local criminal gangs pretending to be police. The capitalist market still operates to provide most goods and services. However, the many governments are destroying the capitalist markets and practical production systems by over-regulating them. The tax code for the Federal Government, not mentioning the tax codes for the 50 states, 300 cities, and thousands of counties, is about 30 million words. No one claims to understand the whole tax code. The Federal regulation, not mentioning state, city, and county regulations, on medical care, healthcare equipment, drugs, health providers, and health insurance organizations is about 100 million words. This is only two industries. The first step toward totalitarianism is endless complex regulations.

Recommended policies to provide governments that provide in accordance with their purpose and no more

All governments need to substantially reduce the number of employees, regulations, debt, and spending fast.

Americans, especially American governments, and in many other countries, need to return to the experience of feeling humility and to experience the feeling of joy in sympathy with others success.

Recommended policies to provide governments that provide in accordance with their purpose: American people must take on the burden of electing honest, intelligent, and vigorous government servants

To have the many good governments that will preserve America and provide for it, American citizens must face the burden of finding honest, intelligent and vigorous candidates for elected office. This is a long and great jump from the current lack of participation with the power that governs.

Individual Americans need to be purified; to return to simpler ways, to obey the rules of decency, to remove the plethora of repressive laws, to subordinate the hundreds of governments to the will of the people. America needs to be ruled by law not by a pretend president, who is an illegal immigrant, not a citizen who stole the election with a huge mob of criminals practicing voter frauds. Americans want to be ruled by virtuous men, not by powerful incompetent elected criminals. These criminals grabbed power because the citizens allowed it.

Americans needs to stop racing after possessions, to stop the craving after prestige and hierarchy, and to stop the clinging to status goods and services.

Americans, Chinese and other large groups of people are afflicted with the same mental disorders: pursuit of power, possessions and prestige. They all stimulate the mental hindrances and ignore the enlightenment factors.

Americans need to return to thinking before acting; not to feeling before acting. Part of the purge, a symptom of decay, is to remove people who are accused of feeling certain ways such as hate and race prejudice. Part of the purge is to remove people who think a certain way. These purges are the shackles of totalitarianism. America needs to throw off the shackles of socialism and totalitarianism. America needs to return to the capitalist principles that made it a great country.

America is suffocating from the lack of spiritual expression, including Christian, Buddhist, Jewish, and Hindu. Christians are being purged for thinking their Bible contains laws to obey. The Christians and Jews are being persecuted for practicing the morals and principles that have been the foundation of countries for thousands of years. American citizens are letting the government and the totalitarians get away with this.

This is the time for the emergence of a system to replace the government, or corporation or religion to take responsibility for providing the basic necessities, goods, services and education. The democratic system is not working neither is the republic system, nor the monarchy, nor the dictatorship, nor is mob rule, nor is anarchy.

Americans need to experiment with systems that are proven to work. Perhaps an emergent system like the siphonophore would gather all the pieces together in a way that they are interdependent but they volunteer to coagulate, to work together for the benefit of all citizens.

Another system that has been successful in a long lived sustained group is the twelve steps and twelve traditions of Alcoholics Anonymous (AA). The AA system allows a flexible administration of power based on the people present in the group and the problems introduced by those people.

The dominating political interests in America are based on wishful thinking, on denial of apparent reality, and the ideas of Marx. Most of the emphasis for most of the many governments in America is based on denying apparent reality which was discussed as the discovery of

Benjamin Schwartz above. The idealistic policies based on denial are massive welfare as a way to get votes, on armed force solutions with America policing the world as the American Empire, and spending borrowed money to enforce the vision of the totalitarian elite class. These policies are destroying the founding principles of America including the Constitution on which all American governments were based at one time.

However, the problem of government in America is also being probed for working solutions, not just idealistic wishes. Frequently, new groups and political parties are founded and their principles are tried out. There is a particularly vigorous approach to persuading society to adopt policies by a plethora of interest groups in San Francisco for the last 50 years beginning with the "hippies" in about 1964. This sincere attempt to introduce special interests into the agenda of the San Francisco population has been accelerating. The recent methods and desired policies of various groups are summarized in the *New Yorker* magazine. (Heller, 2014)

Appendix G: The Legend of Li Bai Who Drowned in the Lake Clinging to the Reflection of the Moon

This is an example of perceiving an event composed of simultaneous levels of physical reality.

Biography of Li Bai

Li Bai (701–762), also known as **Li Po**, was a Chinese poet acclaimed from his own day to the present as a genius and a romantic figure who took traditional poetic forms to new heights. He was one of the two most prominent figures in the flourishing of poetry in the mid-Tang Dynasty that is often called the Golden Age of China.

Around a thousand poems attributed to him exist, thirty-four in the canonical 18th-century anthology Three Hundred Tang Poems. The poems were models for celebrating the pleasures of friendship, the depth of nature, solitude, and the joys of drinking wine. Among the most famous are "Waking from Drunkenness on a Spring Day," "The Hard Road to Shu," and "Quiet Night Thought," which appear frequently in school texts in China today. Legend holds that Li drowned when he jumped from his boat to grasp the moon's reflection in the river.

While Li Bai's mother was pregnant with him, she had a dream of a great white star falling from heaven. This seems to have contributed to the idea of his being a banished immortal, one of his nicknames.

At Chang'an

The Emperor Xuanzong, also known as Emperor Minghuang, summoned Li to the court in Chang'an. Li's personality fascinated the aristocrats and common people alike, who bestowed upon him the nickname "the Transcendent dismissed from the Heaven" or "Immortal Exiled from Heaven." When the emperor ordered Li Bai to the palace, he was often drunk, but quite capable of performing on the spot.

Li Bai wrote several poems about the Emperor's beautiful and beloved Yang Guifei, the highest ranked royal consort. A story, probably apocryphal, circulates about Li Bai during this period. Yang Guifei was persuaded to take offense at Li's poems concerning her. At the persuasion of Yang Guifei, Emperor Xuanzong reluctantly, but politely, and with large gifts of gold and silver, sent Li Bai away from the royal court.

Death of Li Bai

There is a long and sometimes fanciful tradition regarding his death, from uncertain sources, that Li Bai drowned after falling from his boat when he tried to embrace the reflection of the moon in the Yangtze River, Nevertheless, the legend that Li Bai died trying to embrace the reflection of the moon has entered Chinese culture.

The frame of the story in which Li Bai becomes legend

The two lovers sit in the dawn waiting for the sun to show over the mountain. He asks, "Can you find the story of the poet who died from drowning in the lake while he was in love with the reflection of the moon in the water?"

Xingang is delighted to remember, "Honey, I think the poet you mentioned is Li Bai. He is a very famous poet in Tang Dynasty. He initiated the style of romanticism. People call him poetic genius and winebibber. He is a straight-forward poet, making friends by poems and drinking to his heart's desires without control. To Chinese people, suffering is synonymous with life. He came too close to the Emperor. He came to love Yang Guifei. Thus, the emperor exiled him away from his lover."

"In exile, one night, he drinks in his house, getting drunk. With tears covering his face, he walks out of his room and comes upon the lake. Such a beautiful moon in the water! He sees the purity and beauty of it. He wants to believe he can be independent of normal life. He walks toward the moon, unfortunately, he walks into the lake and could never come out again?"

"Let us imagine what happened. In our minds we can have any legend we want. Here is the way I imagine it."

"Yang Guifei is madly in love with Li Bai. She keeps the secret that she is the highest ranking consort of the Emperor. But she uses the name Xinyang. He is hopelessly in love with her. But he does not know her high rank. Xinyang praises Li Bai, 'Thank you so much for your attention! To be honest, in my sick days, I needed you very much. My mind was all about you honey, and in my dream, I seem to have you by my side. Yes, honey, I need health, for the name of our love, I should achieve it. Maybe I haven't paid enough attention to my health before and that's why I was sick. This time it's a cold or fever, but next time, I know it's not just a fever and cold any more. Honey, so nice of you to remind me to take the potions and I will do them one by one according to your great advice!'" "Li Bai runs his hands through her hair and kisses her neck while he moans and snuggles close to her."

Back to Xingang and her lover who is exuberant to hear the story. He encourages her, "If you tell me when you are writing, the hours of which day you are writing, I will enter samadhi and visit your mind. I will try to transmit love into you to fill your heart with satisfaction and you will feel the fulfillment very few women ever feel. That is how much I love you; the power of love enough to send love and vitality to you from 10,000 miles."

Xingang continues, filled with the sexual excitement now that she knows her lover is so enthralled to her.

"As Li Bai stumbles toward the lake, he fades out of his misery remembering the first time he met Xinyang, "Dear Xinyang, you are so gorgeous!"

Xinyang, "Thank you so much."

Li Bai, "I am so surprised to see a goddess of a woman so impossibly beautiful."

Xinyang, "You are so sweet. hahaha, romantic man, there must be so many ladies chase you."

Li Bai, "hahaha, dear I am single my lady. I think I am in love now."

Xinyang, "Wow! That's my honor, in love with me?"

Li Bai, "More than in love, in lust, I must have you! Excuse me for being so direct, I cannot help myself."

Xinyang, "Dear that's fine. I am so happy, can you feel that?"

Li Bai, "I can feel your soft magic. Do you have a soft and lovely feeling now? Do you have a good imagination?"

Xinyang, "Yes! Yes!"

LiBai, "Let us imagine we are in love. What shall we do now my beloved? I am so excited that I am feeling flashes·like lightening througout my body. What are you feeling?"

Xinyang, "I feel happy, so sweet. You are so passionate."

Li Bai, "I am lightly stroking your left foot, can you feel it?"

Xinyang, "Yesssssssss. I feel it now."

Li Bai, "May I kiss your left toes? I am kissing your toes. May I suck your biggest toe on your left foot?"

Xinyang, "Dear! You love feet?"

Li Bai, "Your feet are so tiny and soft. How does this feel to have your toe sucked? Do you like this game?"

Xinyang, "Dear! I have never imagined it!"

LiBai, "If you imagine it, you can have anything you want in your mind."

Xinyang, "Dear, you make me so happy."

Li Bai, "What would you like to have now? You can have anything."

Xinyang, "I want a romantic surprise."

Li Bai, "I will lick your leg from the toe to your thigh. Are you wearing panties?"

Xinyang, "Yes dear, I wear panties."

Li Bai, "May I lick your panties around your love flower?"

Xinyang, "Oh dear! That's so exciting. You will melt me."

Li Bai, "I am licking around your panties. I am sucking your panties."

Xinyang, "Dear, please be tender."

Li Bai, "You are wet around your love flower."

Xinyang, "Oh yes, dripping wet."

Li Bai, "I am rubbing your thighs with my magical hands."

Xinyang, "Oh dear! I want more!"

Li Bai, "I am lifting your love flower to my mouth. I am sliding my hands under your buttocks to lift you up. I am gripping your buttocks and massaging them."

Xinyang, "Dear! My vagina feels empty. I want more! I want you! I want to scream!"

Returning to the lovers discussing Li Bai, we realize that Xingang had used a name similar to her own name instead of the Consort's name. This was a subtle way to put her lover into the mood to seduce her.

Her lover broke into the narrative. "Yes, tell me about him. Please!"

Xingang, continued, "Li Bai is exiled from his hometown during Mid Autumn Festival so he misses the traditional family reunion. But he is torn by loss of Yang Guifei, whom he knew as Xinyang. I myself, have feelings like his as though I were he who is feeling them. I am able to feel his body feeling drunk and falling down and falling asleep. I am feeling my own feelings of love. My feelings are his feelings for his forbidden lover far away in the Capitol."

"In this heart broken mood, he wrote the most famous poem in Chinese history. The poem, "Quiet Night Thought." It runs thus:

靜夜思 A Quiet Night Thought
床前明月光 In front of my bed, there is bright moonlight.
疑是地上霜 It appears to be frost on the ground.
舉頭望明月 I lift my head and gaze at the August Moon,
低頭思故鄉 I lower my head and think of my hometown.

The trip to the place of exile

Xingang recalls, "As Li Bai was walking into exile, he thought of his love far away. "She is mourning and thinking. 'Honey, thank you for making me a fulfilled woman, I am so honored! Your love has great power. You push me to a world of purity and goodness. I feel that I am not living in this Emperor's realm but in an unknown time, maybe in the lost history or the far future. I don't know when it is, but only know that my beloved Li Bai brings me here. He makes a new world for me to inhabit, from the outside to the inner world.'"

"Li Bai thinks, 'She could be my ancient or modern princess forever.'"

"Li Bai is flying to his lover in the spirit world. 'I love you so much that I am you and everything that happens to you also happens to me. I cannot bear the thought that something bad could happen to you. I think I would also have a serious failure of health.'"

"Looking back at the Chang'an city, the palace is as gorgeous as before. The immortal beauties sing and dance in harmony with the good times. Soon they will be dead and the remorse of the Emperor will initiate Yang Guifei into the temple of dead immortal beauties. Li Bai will be the most famous poet in Chinese history because he was able to live in his own reality created in his mind."

"Li Bai felt the deprivation for letting his love slip away. 'It seems that everything is the same after this political fight. The Emperor and his imperial concubines must be drenched in the music and dancing in the royal palace. Will they think of me? Will they think of the past and all the poems I wrote for them? And all people sing high praise for me? No, they will not remember me any more. Chang'an has no room for me at all! I have become a part of history in their laughter!"

"All right! Leaving this palace is a better option for me. Here, I have too many good memories and too little will power to leave. Imagine the previous years, I was the most popular person in front of the emperor, and everyone envied me. I enjoyed the privilege to do almost anything, even the highest ranked Consort was honored to have my poems. But so what? Now, time and trend has changed, I am only an exiled prisoner, not the previous poet at all."

"I am not willing to leave, but I have to leave. The glory here doesn't belong to me any more. But how could I forget everything here? Taking off the first-class clothes, I wear this merry muffler, my heart is cold and depressed by ice. The bad smell of these dirty clothes, the tousled hair and this bleak scene make my tears fill my eyes. I look for her among people, for her pair of eyes. Will she see me off at this moment? Maybe it's the last time I will see her. No, how could I meet her in these worn-out, weary clothes? I feel so ashamed and guilty to see her! I should lower down my head so that no one could recognize me. I walk in a hurry but my heart doesn't want to leave a all!"

"I hear my stomach calling for food! I am so ashamed on this street! I beat my belly to stop the sound but it seems to be a naughty baby, teasing me all the time. Yes! It suddenly occurred to me that I haven't eaten anything for two days! Yes, I am hungry but my spirit is hungrier than my body. Once I have lost the spiritual support, I am just a dying body without a soul. Only my mind is living weakly, and only the memory is existing there."

"Look! The night comes! In the moonlight, such beautiful memories I have! Her face and slim body appear in my mind again. No! She has never left; she is in my mind all the time! Every lonely night, I feel her body, just next to me, smooth skin, soft body and warm heart. I almost forgot how I kissed her body little by little, even how she moaned my name loudly. That beautiful moment makes me cry now. But I dare not cry. I try to keep myself calm, Yes, I keep my crying back, but tears comes out. Is that tears? I don't know, I just feel something wet run around my face, falling down slowly.

The setting of the tragedy

At the beginning of the tragedy, there are sensations. There is the smell of the lake, over grown with lotus blossoms and trotting debris. There is the taste of the country wine in his mouth. There is the appearance of the rotting grass mats on the floor of the cottage. There is the touch on his skin of dirty clothes, the force of the wind, the discomfort of high humidity. His memories of hunger due to drinking wine, not eating anything yesterday and today. His memories of his love, Xinyang, far away. His day in Court when he was sentenced to exile. His memory of his Xinyang, strangled to prove the power of a Court official, who he knew he would never see again. His feelings of wholeness when she was near him and when they made love for hours on end all night. The hours of his fulfillment, his fullness, his massive love pouring forth all night as they fucked one another in 100 positions expressing their emotional. love.

That is the setting.

Li Bai feels remorse as he calls out to is lover who is dead, far away

After having a shower, he lie in bed, read their letters, saw their pictures drawn in the overwhelming weather of love. He missed her and then brought her his evening kiss. Thinking through the haze of his drunken mind, "Honey, what are you doing now? Missing me, just as I how I miss you? My body is in exile, but my heart has found you in the halls of the dead. If time could test my heart for you, then I will stay with it and be your loyal friend immediately. If distance could tell how much I love you, then I will become an angel, fly up to you although you may not see me. If words could express how much I miss you, then I would make the most powerful words, with heart and life but now, I cannot find a suitable way to express my feelings! Every time, I write you in the valley of death, I want to write everything, but I also want to write nothing! My everything is about my daily life and work, wish you could feel it together with me; my nothing is my pure heart, only you, no other woman could enter. Honey, where are you? if I ask you to kiss me now, will you come to me softly? if I want to make you as my only woman, will you say yes and kiss my hand? I know you will, although you are distant from me now, but I know you are with me, never left, never will leave! Now, just come to me and hug your man! He is waiting for you. Maybe you could see these words written on the clouds, but you cannot see the tears that flow for you, when I miss you deeply. All my love Li Po."

Suddenly, optimism fills him as he joyfully remembers. His darling feels real and present, sublime as he embraces her.

In his love drunk mind, she says, "Do you like sunrise? When you see the beautiful sunrise in the morning, what will you think of? Is it beautiful? To be honest, I had another name when I was young, but now, seldom people know it, except my close relative. It's Xinyang. Do you know the meaning of this name? Morning sunrise. Morning sunrise is a new day's hope, when we see it, we see hope in life. More poetically, it stands for a promise in life. The promise for love and life, once you achieve it, hold it tightly for the rest of your life, or until the sunset."

"Honey, you are my promise in life. Could I be your sunrise to bring you hope and warmth? From now on, you could call me Xinyang. I will like it. Just imagine a scene, we stand on the mountain top, waiting for the sun to rise. We sit there hand in hand, whispering and smiling. We look at the sky, excitement and hope are in our hearts. We think nothing at that moment, but the world consists only of we two. Finally, we see it rising little by little. We look into each other's eyes in this beautiful sunrise, we kiss each other. This one moment we remember for the rest of our short lives."

The rising feelings of the story

The two lovers pause in their pondering and affected remorse. Xingang narrates, "I will go to Li Bai's inner world little by little and as your guide. I will explain it little by little."

"I feel like myself living in the Tang dynasty, a handsome scholar. Fallen in love with a pretty girl who I did not know was an immortal beauty. He was drenched in the poetic craft and writing skills! The night, the moonlight, the poet, the missing of home town, all these are alive again! Thank you so much for the inspiring me to guide you! You are my teacher."

Depressed Li Bai lying in the decaying shack

Li Bai pouts, pitying himself, "I failed to control my tears now. People often say, real men do not easily cry, but they just haven't touched the saddest place! I am only an ordinary person, having the seven emotions and six sensory pleasures. Now, look at this wet, narrow, dull and lonely room, only my heart is still alive, and only through my heart beating could I feel that I am living in this world."

"The green bamboo bed has faded in the suffering of years. The gray and moldy smell sickens my nose. Suddenly I feel everything in this room has been here for thousands of years. I will die if I am one more minute here in this room."

"Walking out, the crickets are singing, or are they crying? Dear crickets, are you looking for your love? Or are you lost in this endless night and can't find the way back home? When young, I often caught them for fun to feed the chicken, or duck. Now, I listen to their crying. I know they have lost their only love. Were their loves caught by the naughty kids or missing on the half way? The frogs also come to join. Now, it must be harvesting on the farm; the peanuts, the corns, and the beans. Also, I caught the frogs for fun in my childhood but now, every sound of them makes my body hair stand up straight! They seem to read my heart and cry for me. They seem to call my distant relatives, and they seem to be pitying me. Even the frog and crickets could listen to my heart, but where are you beloved Xinyang?"

Li Bai stumbles to the lake, he feels grateful to his lost lover for the memories, the excruciating pain of lost love. "You have given me such feelings. One of my feelings is great love for you, unknown in the center of China among the lost ancient capitals of imperial China."

"In the fuzzy night, a boat seems to wait for my travel around the lake. I get on it, little by little, go toward the lake center. Look at this gentle water waves, making me shake from side to side, just like her hands, shake me in a swing. With the cool wind, I row little by little. Oh, look at this moon in the water! It runs after me! Are you calling me back, my love? Are you waiting for me in the peach orchard? Yes! It must be you. Now, you come to find me. I can't wait to hug you. You seem to be so beautiful in the water! I want to kiss you. I want to touch you! I move my body and then I feel your cool body. Are you cold my love? No hurry, I will warm you with my body. I will be cold together with you. I almost lost myself in your love. Now I am left with just the cool wind but so comfortable. Am I dreaming? No! I feel I live in the real world. I create my own reality as I do in my poems."

"Li Bai felt remorse, 'Mid Autumn Festival is such a happy time for a family reunion! At this moment, all my family and relatives shall be talking and laughing in the beautiful moonlight. I seem to see the ray of light reflected on their faces. I seem to taste their delicious dishes cooked by kind elders. I seem to hear that someone mentioned my name. I want to propose a toast to them. I want to bring them my blessings from a far distance here, but my voice is so powerless. Before my toast finds them, it has been swallowed by the cold night, by the strong wind, by the mountains, and rivers between us.'"

"Could anyone remember to toast me back in this reunion time? No matter what, I hold my wine cup, come to my window, toast to the direction of my hometown. My spirit is melting because I am missing and loving you all in my cup of wine! In the moonlight, I see my single thin figure, only my figure and the moonlight! I give myself a forced smile, all right, moonlight should be enough in this lonely night, but why do I still feel sad and disappointed? One cup of wine after another. My vision becomes fuzzy, my body becomes shaky. I walk out of my room slowly, little by little. Listen! Laughter of a girl comes out from the neighbor's wall. She must be enjoying this romantic gathering with her lover. But where's mine? I happened to meet you in the peach orchard. You are inhaling the fragrance of the sweet flower, and smiling gently! But how could I forget your slender waist, your sweet smile, your soft voice, and your smooth long hair?"

"Little by little, I come to this beautiful lake. Look at this lake of water! Do they know what is missing? Do they know the magic and depression of love? The lake must not know my feelings now. How could there be safety and security? Without any wave, without any sound, the lake is alive with unseen animals. Suddenly the round moon reflected in the lake attracts my eyes! I see a beautiful figure in it. Is that Xinyang? I guess so. I walk to it without control. I left my shoes, but don't want to pick them. I seem to touch her but suddenly, the moon breaks into pieces in the water. Where am I? I have no consciousness at all; only this quiet night; so quiet; without any sound. I seem to hear something calling to me, but also seem to hear only the thinking of Xinyang; the delusions of this quiet lonely night.

The climax

Li Bai cheers up realizing his true love is with him, "I sit in the boat, smelling the fungus and the rotten wood of the boat; hearing the tiny waves hitting the boat. Suddenly, I see the reflection of the moon. The Chi rises in my backbone like a dragon. I am invigorated. The moon looks like Xinyang! I can see her face on the moon!

I am overwhelmed by love and lust for my darling. It is so perfect that she is on the moon. Now the moon has come to me. She is in the lake. I am enthralled by the sight of her. I must have you my love! NOW we will make love again. What is this thing in the way? The railing. Get out of the way railing! I have a date with the light of my life. The water is so warm. I am bathing with my sweetheart. It is so soft and pure. I embrace you, the mother of my children. There she is floating in the lake. Wait for me! I am coming! I am kissing you! I do not need to breathe. I must have you! I cannot stop! I feel the bump on my head. The boat again! Get out of the way boat! Now everything is fading out as I merge with my goddess. I am drowning in the lake while I embrace the reflection of the moon. It does not matter!

The lovers blissfully comment on the story

He is crying. Through the tears running over his lips, "Dearly beloved Xingang, precious product of 10,000 years of Chinese excellence, I am so sad for Li Bai that tears roll down my cheeks. I must hold my back my sobs, so I can speak to you, my great beloved. I do not know why your writing is so sad and it transmits such heavy tragedy. But you are splendid at this type of tragic writing. I am so affected I want to eat your whole body to contain such genius."

Xinyang and her lover embrace while both exploding into sobs. The tears cover their clothes and fill their shoes.

He continues, "Shivers run up and down my whole body. I need to hold you in my arms and to lick your whole body to taste the excitement of your excellent understanding the human heart. How can I express my wonder at your growing ability to tell such moving stories? The thrill bounces around in my body from the back of my head to my stomach to my sexual organ."

"I have to rest to bring back my reasonable scientific nature before I can continue. I love your responsiveness. You are so alive in this story telling. I still want to consume you. I have to wait until my hands stop shaking. The impression of Li Bai is so great!"

"You told me what Li Bai thinks in his drunken mind, tastes, touches his bamboo wall, smells his moldy wet bamboo floor, hears the crickets and the frogs by the lake. Then his advance toward the climax: walking to the edge of the lake, getting into his boat, feeling love for the moon as though she were his lost love. Then the climax, which is like a sexual orgasm: He cries out to the moon and grabs her, falling into the lake and not caring that he is wet and helpless. Feeling the suffocation of the water. Then the end: the last thing he hears is the water waves as his head bumps the bottom of the boat. You light up the lives of every Chinese through the death of a great poet. Chinese have heard of him so the rest of the story is already in their memories."

"The delusion of Li Bai already is part of Chinese minds who have been reading his poems for 1300 years. The combination of the mind ready to believe and the colorful thinking within the story telling creates a new reality."

Her lover exclaims to Xingang, "You are blessed by your past lives. You must have been his lover in a past life. You have such a mellifluous feeling for him."

The ending of the frame in which Li Bai became legend

Xingang recalled, "You must know the famous poet William Butler Yeats in Ireland, and you must know his wide-spread poem:" *When you are old.* True love could go through time and space to reach the deep inner heart and make a great legend! When love doesn't care for appearance and age, it finds its home and destiny. Look at this beautiful poem,

> When you are old and grey and full of sleep
> And nodding by the fire, take down this book
> And slowly read, and dream of the soft look
> Your eyes had once, and of their shadows deep;
> How many loved your moments of glad grace,
> And loved your beauty with love false or true,

But one man loved the pilgrim soul in you,
And loved the sorrows of your changing face;
And bending down beside the glowing bars,
Murmur, a little sadly, how love fled
And paced upon the mountains overhead
And hid his face amid a crowd of stars.

Her lover, crying without control, "You wrote this while you were in the seventh level in the heaven of love. Then you gave 10,000 Chinese the definition of love which they can use to measure the decision to marry. You will make me cry a river. And you will live for 10,000 years in the childrens' books.

Other views of the legend of Li Bai

The all powerful and invincible Dragon
sails through the enchanting clouds
which obscure Heaven.
The arrogant, cunning and wise Tiger
sneaks through the mists which
hide Earth.
Heaven and Earth totally control Man
until the clouds and mists disappear
then a mosquito and a cockroach
are all that remain.
A monk recalled the Buddha's description of the ineluctable essence of the source of reality in the *Diamond Sutra*.
Thus have I heard:
"This is the correct view when contemplating our conditioned existence in this fleeting world:
Like a shimmering star, a flickering lamp,
Like a fleeting Autumn cloud, a flash of summer lightening,
Like a tiny shining drop of morning dew, a bubble floating in a stream;
Like an illusion, a phantom, or a dream.
So is all conditioned existence to be seen."

References

Acocella, Joan (2013) "Misery: Is there justice in the Book of Job? New York: *The New Yorker*, December 16, 2013, pp.83-87.

Acocella, Joan (2014). "Selfie: how big a problem is narcissism," New York: *The New Yorker*, May 12, 2014, pp. 77-81.

Adair, Robert Kemp (1987). *The Great Design: Particles, Fields and Creation*, New York: Oxford Univ. Press, Chapter 8.

Anonymous (1910a). "Kings I," *The Holy Bible*, New York: The Cambridge Univ. Press, Chapter 11: 4-11

Anonymous (1910b) "Job," *The Holy Bible*, New York: The Cambridge Univ. Press.

Anonymous (1910d) "Jonah," *The Holy Bible*, New York: The Cambridge Univ. Press.

Aristotle (1936). *Minor Works: On Colors. On Things Heard. Physiognomics. On Plants. On Marvellous Things Heard. Mechanical Problems. On Indivisible Lines...*, W. S. Hett, trans. Gorgias (Loeb Classical Library No. 307), *Physiognomics* 808b11 Chapter IV.

Ash, Robert B. (1965). *Information Theory*, New York: Dover.

Astola, Jaakko (1997). *Fundamentals of nonlinear digital filtering*, Boca Raton, FL: CRC Press.

Bird, Richard J. (2003). *Chaos and Life: Complexity and Order in Evolution and Thought*, New York: Columbia Univ. Press.

Auerbach, Erich (2003). *Mimesis: The Representation of Reality in Western Literature*, 50th English anniversary ed., Willard R. Trask, trans., Princeton, NJ: Princeton Univ. Press.

Barrow, John D. (2009). *The Book of Nothing: Vacuums, Voids, and the Latest Ideas about the Origins of the Universe*, New York: Vintage Books.

Buddha (500BC). "10 Satipatthana Sutta: The Foundations of Mindfulness," *The Middle Length Discourses of the Buddha: A New Translation of the Majjhima Nikaya*, Bhikkhu Nanamoli and Bhikkhu Bodhi, trans., Boston: Wisdom Pub.

Calvocoressi, Peter (1980). *Top Secret Ultra*, New York: Pantheon, p. 42ff.

Chernoff, Herman and Moses, Lincoln E. (1959). *Elementary Decision Theory*, New York: Dover.

Coulson, C. A. (1977) *Waves: A Mathematical Approach to the Common types of Wave Motion*, New York: Longman.

Dalai Lama (2012). *From Here to Enlightenment: An Introduction to Tsong-Kha-Pa's Classic Text the Great Treatise on the Stages of the Path to Enlightenment*, Boston: Snow Lion.

Denbigh, K. G. (1951). *The Thermodynamics of the Steady State*, London: Methuen, and Hoboken, NJ: Wiley.

Des Pres, Terrence (1976). *The Survivor: An Anatomy of Life in the Death Camps*, New York: Oxford Univ. Press.

Dhammajothi, Mădavacciyē, Himi and Dhammajothi, Medawachchiye Thero (2009). *Concept of Emptiness in Pali Literature*, Colombo: International Publishers.

Dodgen, Randall A. (2001.) *Controlling the Dragon: Confucian Engineers and the Yellow River in late Imperial China*, Honolulu, HI: Univ. Hawaii Press.

Einstein, Albert (1904). *AdP*, Vol. 4, p. 354.

Einstein, Albert (1909). *Phys. Zeitschr.*, Vol. 10, p. 185.

Fyfe, Albert J. (1988). *Understanding the First World War: illusions and realities*, New York: P. Lang.

Gaylin, Willard (2003). "Chapter 11 Identifying the Enemy," *Hatred: The Psychological Descent into Violence*, New York: Public Affairs, p. 173ff.

Genz, Henning (1999). *Nothingness: the Science of Empty Space*, Reading, MA: Helix books imprint of Perseus.

Goodman, J. W. (1968). *Introduction to Fourier Optics*, New York: McGraw-Hill.

Graham, Benjamin and Dodd, David L. (1934). *Security Analysis*, New York: Whittlesey House.

Gulrajani, Ramesh M. (1998) *Bioelectricity and BioMagnetism*, Hoboken, NJ: Wiley.

Heidegger, Martin, (1962). *Being and Time,* 7th ed. John Macquarrie and Edward Robinson trans. New York: Harper & Row.

Heisenberg, Werner (1962) *Physics and Philosophy: the Revolution in Modern Physics*, New York: Harper Torch imprint of Harper & Row.

Heller, Nathan (2014). "Letter from San Francisco: California Screaming: the tech industry made the Bay area rich. Why do so many residents hate it?" *New Yorker*, July 7&14, 2014, pp.46-53.

Hochberg, Julian E. (1978). *Perception*, second edition, Englewood Cliffs, NJ: Prentice-Hall.

Hochschild, Adam (2011). *To End All Wars: A Story of Loyalty and Rebellion, 1914-1918*, New York: Houghton, Mifflin, Harcourt.

Hodgkin, A. L., and Huxley, A. F., "A quantitative description of membrane current and its application to conduction and excitation in nerve." *J. Physiol.* (1952) 177, 500-544

Howard, Robert G. (2012a). "Chapter 9 The Ultimate Ground of Human Experience" *Mind. Consciousness, Body: Hypothetical and Mathematical Description of Mind and Consciousness Emerging from the Nervous System and Body*, Bloomington, IN: iUniverse.

Howard, Robert G. (2012b). "Chapter 11 The Mathematical Theory of Consciousness, *Mind, and Time*" *Mind. Consciousness, Body: Hypothetical and Mathematical Description of Mind and Consciousness Emerging from the Nervous System and Body*, Bloomington, IN: iUniverse.

Howard, Robert G. (2012c). "Chapter 13 Derivation of the Conservation Equation for Content of Non-conscious Mind: the basis of Diffusion Mathematics Describing the Mind" and

"Chapter 14 Description of Diffusion of Non-conscious Contents into Conscious Mind" and

"Chapter 15 Diffusion of all factors of NCC and conscious contents based on entropy flow and information theory" *Mind. Consciousness, Body: Hypothetical and Mathematical Description of Mind and Consciousness Emerging from the Nervous System and Body*, Bloomington, IN: iUniverse.

Howard, Robert G. (2012d). "Chapter 7 "Discovering the Laws of Psychic Science," *Mind. Consciousness, Body: Hypothetical and Mathematical Description of Mind and Consciousness Emerging from the Nervous System and Body*, Bloomington, IN: iUniverse.

Howard, Robert G. (2012e). "Chapter 4 The Geometry of Precognition, Chapter 5 Time to Formulate the Laws and Hypotheses of Psychic Science, Chapter 6 Discovering the Hypotheses of Psychic Science, Chapter 7 Discovering the Laws of Psychic Science, Chapter 8 Can an Ordinary Person Be Trained to Use the Psychic Senses?" Mind. Consciousness, Body: Hypothetical and Mathematical Description of Mind and Consciousness Emerging from the Nervous System and Body, Bloomington, IN: iUniverse.

Hynes, Samuel (1997). *The Soldier's Tale: Bearing Witness to Modern War*, New York: Penguin.

Hull, John (1989). *Options Futures and other Derivative Securities*, Englewood Cliffs, NJ: Prentice Hall.

Hutschnecker, Arnold A, (1974). *The Drive for Power*, New York: Evans.

Jackall, Robert (1988). *Moral Mazes: The World of Corporate Managers*, New York: Oxford Univ. Press.

Johnson, David E. (1976). *Introduction to Filter Theory*, Englewood Cliffs, NJ: Prentice-Hall.

Jung, C. G. (1955). "Synchronicity, an Acausal Connecting Principle" in Jung, C. G. & Wolfgang Pauli *Interpretation of Nature and the Psyche*, R. F. C. Hull, trans., London: Routledge & Kegan Paul.

Jung, C. G. (1969). "On Synchronicity" *Collected Works, Vol. 8, The Structure and Dynamics of the Psyche*, 2nd ed., London: Routledge & Kegan Paul.

Kennedy, Daniel David and Grandmaster Lin Yun (2011). *Feng Shui for Dummies*, Hoboken, NJ: Wiley.

Kennedy, Paul M. (1987). *The Rise and Fall of the Great Powers: Economic Change and Military Conflict from 1500 to 2000*, New York: Random House.

Kohn, Livia and Michael LaFargue, eds., *Lao-Tzu and the Tao-Te-Ching*, Albany: State University of New York Press, 1998.

Kruger, Rayne, (2003). *All Under Heaven: A Complete History of China*, The Atrium, Southern Gate, Chichester, West Sussex, England: Wiley, pp. 64-85.

Kuhn, Thomas S. (1970) *The Structure of Scientific Revolutions*, 2nd ed. Chicago: Chicago Univ. Press.

Lanchester, John (2014b). "Money Talks: Learning the language of finance," *The New Yorker*, August 4, 2014, pp. 30-33.

Lanchester, John (2014a). "Shut Up and Eat: A Foodie Repents," *The New Yorker*, November 3, 2014, pp36-38.

Levin, M. R. (2012). *Ameritopia: The Unmaking of America*, New York: Threshold.

Liu, Xie (1959). *The Literary Mind and the Carving of Dragons*, Translation of Wen-hsin tiao-lung, New York: Columbia University Press.

Mann, Thomas (1964). *The Magic Mountain*, H. T. Lowe-Porter, trans., New York: Knopf, 1964, pp. 393-410) First Ed. 1924

Maxwell, James Clerk (1954). *A Treatise on Electricity and Magnetism*, 3rd ed. New York: Dover.

Mikhailov, Mikhail Dimitrov and M. Necati Ozisik (1984) *Unified Analysis and Solutions of Heat and Mass Diffusion*, Wiley, New York.

Moore, Michael S. (1986). *A Natural Law Theory of Interpretation*, The Semantics of Judging, New York: Oxford Univ.Press.

Mountcastle, V. B., (1975). "The View from within: Pathways to the Study of Perception" *Johns Hopkins Medical Journal*, 136, p. 109-131.

Page, Jake (2003) *In the Hands of the Great Spirit: the 20,000 Year History of American Indians*, New York: Free Press a division of Simon Schuster.

Pandarakalam, James Paul (April 2011). Medjugorje Apparitional Occurrences: A Para-psychological and Spiritual Analysis, *Journal of Spirituality and Paranormal Studies*, Vol. 34, No. 2, 100-117.

Pandarakalam, James Paul (October 2012). Medjugorje Apparitional Experiences, *Journal for Spiritual and Consciousness Studies*, Vol. 35, No. 4, 196-211.

Paul, Ron (2011). *Liberty Defined:50 Essential Issues that Affect Our Freedom*, New York: GrandCentral.

Peat, F. David, 2002. *From Certainty to Uncertainty: The Story of Science and Ideas in the Twentieth Century*, Washington, DC: Joseph Henry Press.

Pinsky, Mark A. (2008). *Introduction to Fourier Analysis and Wavelets*, Providence, RI: American Mathematical Society.

Plato "Allegory of the Cave" in *The Republic*, Oxford Univ. Press, New York, 1959, pp. 227-231.

Popper, Karl R. and Eccles, John C. (1977). "Chapter E1 The Cerebral Cortex" *The Self and Its Brain: An Argument for Interactionism*, London: Routledge & Kegan Paul.

_____ "Chapter P1, p. 9-10.

_____ "Chapter P2 The Worlds 1, 2, and 3," p. 36-50.

_____ "Chapter P3 Materialism Criticized," p. 51-99.

Prabhavananda, Swami and Isherwood, Christopher (1953). *How to Know God: the Yoga Aphorisms of Patanjali*, Swami Prabhavananda and Christopher Isherwood, trans. Hollywood, CA: Vedanta Press.

Rickards, James (2014). *The Death of Money: The Coming Collapse of the Inernational Monetary System*, New York: Portfolio imprint of Penguin.

Rossbach, Sarah (2000). *Feng Shui: the Art of Chinese Placement*, Penguin, New York.

Salisbuty, Harrison E. (1969). *War between Russia and China*, New York: Norton.

Schwartz, Benjamin (1985). *The World of Thought in Ancient China*, Cambridge, MA. Harvard Univ. Press.

Shannon, C. E. (1949). "A Mathematical Theory of Communication," *Bell System Technical Journal 27*, p. 379-423 and p. 623-656, 1948. Reprinted in C. E. Shannon and W. Weaver, *The Mathematical Theory of Communication*, Urbana, IL: Univ. Illinois Press.

Smith, Gregory D., Charles L. Cox, S. Murray Sherman, John Rinzel (2000). "Fourier Analysis of Sinusoidally Driven Thalamocortical Relay Neurons and a Minimal Integrate-and-Fire-or-Burst Model" *J. Neurophysiol.* 83:588-610, 2000.

Soustelle, Jacques (1970). *Daily Life of the Aztecs: On the Eve of the Spanish Conquest*, Patrick O'Brian, trans., Redwood City, CA: Stanford University Press.

Stein, John F. and Catherine J. Stoodley (2006). *Neuroscience: An Introduction*, New York: Wiley, p.111.

Sun Tzu (1971). *The Art of War*, Samuel B. Griffith trans., Oxford, Oxford Univ. Press.

VanSlyke, Lyman (1988). *Yangtze: Nature History and the River*, New York: Addison-Wesley.

Unknown Author (1988). *The T'aoist Classics: the Collected Translations of Thomas Cleary Vol. 3 Vitality, Energy, Spirit, The Secret of the Golden Flower, Immortal Sisters, Awakening to the Tao*, Boston: Shambala.

Vaĭnshteĭn, Lev Al'bertovich (1962*). Extraction of Signals from Noise*, Richard A. Silverman, trans, Englewood Cliffs, NJ: Prentice-Hall.

Vaseghi, Saeed V. (2006). *Advanced Digital Signal Processing and Noise Reduction*, Hoboken, NJ:Wiley.

Vasquez, Juan Gabriel (2013). *The Sound of Things Falling*, Anne McLean, trans., New York: Riverhead imprint of Penguin. (originally 2011 Spain: Alfaguara).

Vedral, Vlatko (2010). *Decoding Reality: The Universe as Quantum Information*, New York: Oxford Univ. Press.

Waley, Arthur (1938). *The Analects*, London: George Allen and Unwin. Reprinted by Alfred A. Knopf in 2000 ISBN 978-0-375-41204-2).

Watts, Alan, 2000. *What is Tao?* Novato, CA: New World Library.

Wiener, Norbert (1961). "Time Series, Information, and Communication," *Cybernetics or Control and Communication in the Animal and the Machine*, 2d. ed., Cambridge, MA: MIT Press. Pages 60-94

Wigner, Eugene P. (1939). *Ann. Math.*, Vol. 40, p. 149.

Wigner, Eugene P. (1967). *Symmetries and reflections: Scientific essays of Eugene P. Wigner*, Bloomington, IN: Indiana Univ. Press.

Williams, Arthur Bernard and Taylor, Fred J. (1995). *Electronic filter design handbook*, 3rd ed., New York: McGraw-Hill.

Wilson, Raymond, G. (1995). *Fourier Series and Optical Transform Techniques in Contemporary Optics*, New York: Wiley

Zha, Jianying (2011) *Tide Players: The Movers and Shakers of a Rising China*, New York: New Press.